Handbook of Compliant Mechanisms

Handbook of Compliant Mechanisms

Edited by

Larry L. Howell
Brigham Young University, USA

Spencer P. Magleby
Brigham Young University, USA

Brian M. Olsen
Los Alamos National Laboratories, USA

A John Wiley & Sons, Ltd., Publication

Library of Congress Cataloging-in-Publication Data

Handbook of compliant mechanisms / edited by Professor Larry Howell, Dr. Spencer P. Magleby,
Dr. Brian M. Olsen.
      pages cm
   Includes bibliographical references and index.
   ISBN 978-1-119-95345-6 (cloth)
   1. Mechanical movements.   2. Machinery, Kinematics of.   3. Engineering design.   I. Howell, Larry L.,
editor.   II. Magleby, Spencer P., editor.   III. Olsen, Brian M. (Brian Mark), 1983– editor.
   TJ181.H25 2013
   621.8–dc23

                                                                                        2012037077

Set in 10/12pt Palatino by Aptara Inc., New Delhi, India

Contents

PART THREE   SYNTHESIS OF COMPLIANT MECHANISMS

List of Contributors

Chapter Contributors

Shorya Awtar – Assistant Professor, University of Michigan, Ann Arbor, MI, USA
Mary Frecker – Professor, The Pennsylvania State University, University Park, PA, USA
Jonathan Hopkins – Assistant Professor, University of California, Los Angeles, CA, USA
Larry Howell – Professor, Brigham Young University, Provo, UT, USA
Brian Jensen – Associate Professor, Brigham Young University, Provo, UT, USA
Charles Kim – Assistant Professor, Bucknell University, Lewisburg, PA, USA
Girish Krishnan – Post-Doctoral Associate, University of Michigan, Ann Arbor, MI, USA
Craig Lusk – Associate Professor, University of South Florida, Tampa, FL, USA
Spencer Magleby – Associate Dean, Brigham Young University, Provo, UT, USA
Chris Mattson – Associate Professor, Brigham Young University, Provo, UT, USA
Brian Olsen – Research Engineer, Los Alamos National Lab, Los Alamos, NM, USA

Managing Library Contributors

G. K. Ananthasuresh – Indian Institute of Science, Bangalore, Bangalore, India
Guimin Chen – Xidian University, Xi'an, P.R. China
Martin Culpepper – Massachusetts Institute of Technology, Cambridge, MA, USA
Mohammad Dado – University of Jordan, Amman, Jordan
Haijun Su – Ohio State University, Columbus, OH, USA
Simon Henein – CSEM Centre Suisse d'Electronique et de Microtechnique SA, Neuchâtel, Switzerland
Just L. Herder – Delft University of Technology, Delft, The Netherlands
Jonathan B. Hopkins – University of California, Los Angeles, CA USA
Nilesh D. Mankame – General Motors Research & Development, Warren, MI, USA
Ashok Midha – Missouri University of Science and Technology, Rolla, MO, USA

Anupam Saxena – Indian Institute of Technology, Kanpur, Kanpur, India
Umit Sonmez – American University of Sharjah, Sharjah, UAE
Jingjun Yu – Beihang University, Beijing, China

Library Contributors

Imad F. Bazzi – General Motors Research & Development, Warren, MI, USA
Shusheng Bi – Beihang University, Beijing, China
Ozgur Erdener – Istanbul Technical University, Istanbul, Turkey
Bilin Aksun Güvenç – Okan University, Istanbul, Turkey
Huseyin Kızıl – Istanbul Technical University, Istanbul, Turkey
Xu Pei – Beihang University, Beijing, China
Ahmet Ekrem Sarı – Altinay Robot Technologies, Istanbul, Turkey
Nima Tolou – Delft University of Technology, Delft, The Netherlands
Levent Trabzon – Istanbul Technical University, Istanbul, Turkey
Cem Celal Tutum – Technical University of Denmark, Lyngby, Denmark
Hongzhe Zhao – Beihang University, Beijing, China
Guanghua Zong – Beihang University, Beijing, China
Yörükoğlu, Ahmet – R&D Engineer at Arcelik

Student Contributors

Bapat, Sushrut – Missouri University of Science and Technology
Barg, Matt – Brigham Young University
Berg, Fred van den – Delft University of Technology
Black, Justin – Brigham Young University
Bowen, Landen – Brigham Young University
Bradshaw, Rachel – Brigham Young University
Campbell, Robert – Brigham Young University
Chinta, Vivekananda – Missouri University of Science and Technology
Dario, P. – Scuola Superiore Sant'Anna
Davis, Mark – Brigham Young University
Demirel, Burak – KTH Royal Institute of Technology
Duffield, Luke – Brigham Young University
Dunning, A.G. – Delft University of Technology
Emirler, Mümin Tolga – Istanbul Technical University
Foth, Morgan – Brigham Young University
George, Ryan – Brigham Young University
Güldoğan, Bekir Berk – Istanbul Technical University
Greenberg, Holly – Brigham Young University
Hardy, Garrett – Brigham Young University
Harris, Jeff – Brigham Young University
Howard, Marcel J. – Delft University of Technology

Ivey, Brad – Brigham Young University
Jones, Andrea – Brigham Young University
Jones, Kris – Brigham Young University
Kluit, Lodewijk – Delft University of Technology
Koecher, Michael – Brigham Young University
Koli, Ashish – Missouri University of Science and Technology
Kosa, Ergin – Istanbul Technical University
Kragten, Gert A. – Delft University of Technology
Kuber, Raghvendra – Missouri University of Science and Technology
Lassooij, Jos – Delft University of Technology
McCort, Ashby – Brigham Young University
Morsch, Femke – Delft University of Technology
Morrise, Jacob – Brigham Young University
Pate, Jenny – Brigham Young University
Peterson, Danielle Margaret – Brigham Young University
Ratlamwala, Tahir Abdul Husain – University of Ontario Institute of Technology
Reece, David – Brigham Young University
Samuels, Marina – Brigham Young University
Sanders, Michael – Brigham Young University
Shafiq, Mohammed Taha – American University of Sharjah
Shelley, Dan – Brigham Young University
Shurtz, Tim – Brigham Young University
Simi, Massimiliano – Scuola Superiore Sant'Anna
Skousen, Darrell – Brigham Young University
Solomon, Brad – Brigham Young University
Steutel, Peter – Delft University of Technology
Stubbs, Kevin – Brigham Young University
Tanner, Daniel – Brigham Young University
Tekeş, Ayşe – Istanbul Technical University
Telford, Cody – Brigham Young University
Toone, Nathan – Brigham Young University
Wasley, Nick – Brigham Young University
Wengel, Curt – Brigham Young University
Wilding, Sam – Brigham Young University
Williams, David – Brigham Young University
Wright, Doug – Brigham Young University
Yu, Zhiwei – Beihang University
Zhao, Shanshan – Beihang University
Zirbel, Shannon – Brigham Young University

Acknowledgments

The *Handbook of Compliant Mechanisms* is a result of work by contributors from around the world. Compliant mechanisms experts have authored the chapters in Parts II and III, and many more have contributed entries to the *Library of Compliant Mechanisms* (Part IV of the *Handbook*). The contributions of these individuals are gratefully acknowledged.

We express appreciation to Brian Winder and Jonathan Hopkins for their work on early drafts of the *Library of Compliant Mechanisms*. The graphic design assistance of Jung-Ah Ahn (Jade) and Stephen Jensen are acknowledged, as is the administrative assistance of Danielle Peterson. The format of the *Library of Compliant Mechanisms* was inspired by Ivan I. Artobolevskii's seven-volume work *Mechanisms in Modern Engineering Design: A Handbook for Engineers, Designers, and Inventors*. We also wish to honor the memory of Dr. Umit Sonmez, who passed away unexpectedly during the time that he was contributing to the *Handbook*.

Acknowledgments

Preface

Compliant mechanisms are seeing expanded use because they offer advantages such as increased performance (e.g. high precision, low weight, low friction), lower cost (e.g. simplified manufacture, low part count), and ability to miniaturize (e.g. makes possible micro- and nanomechanical devices). However, because compliant mechanisms are relatively new compared to more traditional devices, it is difficult for designers to find examples and resources to guide them in their work. Many people are beginning to understand the advantages of compliant mechanisms but there is still a general lack of knowledge of how to implement them. Although many journal articles and some texts are available to aid in the in-depth engineering of compliant mechanisms, a more concise and visual resource is needed to provide inspiration and guidance in the conceptual stages of compliant mechanism design.

The *Handbook of Compliant Mechanisms* is intended to provide a summary of compliant mechanism modeling and design methods and a broad compilation of compliant mechanisms that will provide inspiration and guidance to those interested in exploiting the advantages of compliant mechanisms in their designs. Early *Handbook* chapters provide basic background in compliant mechanisms, summaries of some of the major methods for designing compliant mechanisms, categories of compliant mechanisms, and an example of how the *Handbook* can be used to facilitate compliant mechanism design. Graphics and brief descriptions of many compliant mechanisms are provided to give inspiration in preliminary design.

The *Handbook of Compliant Mechanisms* is designed to be a resource for engineers, designers, and others involved in product design. We hope that it is found to be useful by many in the development of compliant mechanisms.

The *Handbook* is divided into the following Parts:

Part I provides an introduction to compliant mechanisms and describes how to use the *Handbook* to design compliant mechanisms.
Part II focuses on modeling of compliant mechanisms.
Part III describes methods for the synthesis of compliant mechanisms.
Part IV is a visual library of compliant mechanisms.

We wish to express our sincere thanks to all the contributors that worked to make this handbook possible. We hope that it is found to be useful in creating new compliant mechanism designs.

This material is based upon work supported by the National Science Foundation under Grant No. CMMI-0800606. Any opinions, findings, and conclusions or recommendations expressed in this material are those of the authors and do not necessarily reflect the views of the National Science Foundation.

Brian M Olsen is an employee of Los Alamos National Security, LLC, the operator of Los Alamos National Laboratory for the US Department of Energy. The views expressed in this book are solely those of Brian and the other authors and do not necessarily reflect the views, positions and opinions of the US Department of Energy or the US Government.

Larry L. Howell
Brigham Young University, USA

Spencer P. Magleby
Brigham Young University, USA

Brian M. Olsen
Los Alamos National Laboratory, USA

November 2012

Part One
Introduction to Compliant Mechanisms

Part One
Introduction to Compliant Mechanisms

1

Introduction to Compliant Mechanisms

Larry L. Howell

Brigham Young University, USA

1.1 What are Compliant Mechanisms?

If something bends to do what it is meant to do, then it is compliant. If the flexibility that allows it to bend also helps it to accomplish something useful, then it is a compliant mechanism [1]. The idea of using compliant mechanisms in products is catching on, but traditionally when designers need a machine that moves, they commonly use very stiff or rigid parts that are connected with hinges (like a door on its hinge or a wheel on an axle) or sliding joints. But when we look at nature we see an entirely different idea from rigid parts connected at joints – most moving things in nature are very flexible instead of stiff, and the motion comes from bending the flexible parts [2]. For example, consider your heart – it is an amazing compliant mechanism that started working before you were born and will work all day every day for your entire life. Think of bee wings, elephant trunks, eels, sea weed, spines, and the blooming of flowers (Figure 1.1) – all of which are compliant. Even the natural motions that seem to be exceptions to this bending behavior, like your knee or elbow, use cartilage, tendons, and muscles to do their work. We see in nature the possibility of making machines that are very compact – a mosquito (Figure 1.1) is able to fly while carrying its own on-board navigation, control, energy harvesting, and reproduction systems. Would it be possible for us to improve human-designed products if we applied the lessons learned from nature and looked to flexibility to achieve movement?

It is interesting that some early man-made machines were compliant mechanisms. Is that because we were closer to nature then? An example of a compliant mechanism with a multi-millennia history is the bow (Figure 1.2). Ancient bows were made using a composite of bone, wood, and tendon, and they used the flexibility of their limbs to

Handbook of Compliant Mechanisms, First Edition. Edited by Larry L. Howell, Spencer P. Magleby and Brian M. Olsen.
© 2013 John Wiley & Sons, Ltd. Published 2013 by John Wiley & Sons, Ltd.

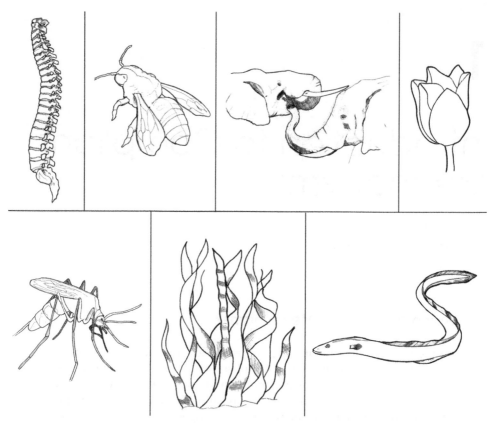

Figure 1.1 A few examples of compliance in nature: a spine, bee wings, elephant trunks, blooming flowers, a mosquito, sea weed, and eels

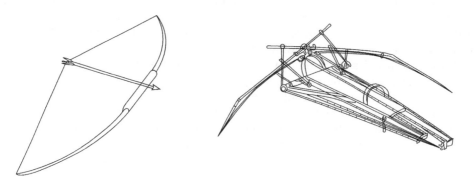

Figure 1.2 Early compliant mechanism designs include the ancient bow and many compliant mechanism designs by Leonardo da Vinci

Figure 1.3 The Wright brothers used wing warping to achieve control of their aircraft for sustained human flight

store energy that would be released into propelling the arrow. It is interesting to see the sketches of Leonardo da Vinci [3] and see many compliant mechanisms (see Figure 1.2 for an example). Even one of the great achievements of engineering – sustained human flight – began with a compliant mechanism when the Wright brothers (Figure 1.3) used wing warping to achieve control of their early aircraft [4].

This may all sound good, but it turns out that compliant mechanisms can be difficult to design. Nature has done it, but nature employed very different design methods from those we mortals use. Great strides were made in the design of machines when compliance was left to nature and we moved to the much easier-to-design realm of rigid parts connected at hinges. For example, the too-sophisticated-for-its-time wing warping of the Wright Flier was eventually replaced by the much-easier-to-work-with control surfaces provided by an aileron pivoting on a hinge.

However, over the past few decades our knowledge has advanced. We have developed new materials, increased our computational capabilities and expanded the ability to design more sophisticated devices. At the same time, society has developed new needs that cannot be easily addressed using traditional mechanisms. This means that there is an increased ability to create compliant mechanisms, and an increased motivation for doing so. As an example, reconsider the example of aircraft control. The Wright Flier started out with wing warping for its control surfaces, but other aircraft quickly moved to approaches using traditional mechanisms. But with the increased computational power available and improved materials that have been developed, researchers are returning to the idea of wing warping to get the advantages, such as reduced weight, that would come from the approach.

One of the things that make traditional design of mechanical components compelling is that designers can separate different functions to be done by different parts, and each part is assigned to do that one function. The blessing and curse of compliant mechanisms is that they integrate different functions into fewer parts. Compliant mechanisms may be able to accomplish complex tasks with very few parts, but they can be much more difficult to design.

1.2 What are the Advantages of Compliant Mechanisms?

The integration of functions into fewer parts leads to compelling advantages for compliant mechanisms. For one, there is a potential for significantly lower costs. This comes from reduced assembly, fewer components to stock, and the possibility of simplified manufacturing (such as fabricating a mechanism from a single mold).

Another advantage is the potential for increased performance. This includes high precision [5, 6] due to reduced wear and reduced or eliminated backlash. The low weight of compliant mechanisms can be useful for shipping and for weight-sensitive applications such as spacecraft. Eliminating the need for lubrication at joints is also a useful performance improvement that is helpful in many applications and environments.

Another category of advantages lies in the ability to miniaturize compliant mechanisms. Microelectromechanical systems (MEMS) for example, are often fabricated from planar layers and compliant mechanisms offer a way to achieve motion with the extreme constraints caused by the resulting geometry (Figure 1.4) [7, 8]. Compliant mechanisms will likely be central to the creation of nanoscale machines.

1.3 What Challenges do Compliant Mechanisms Introduce?

While the advantages of compliant mechanisms are amazing, they also have some challenges that have to be carefully considered in their design. For example, the

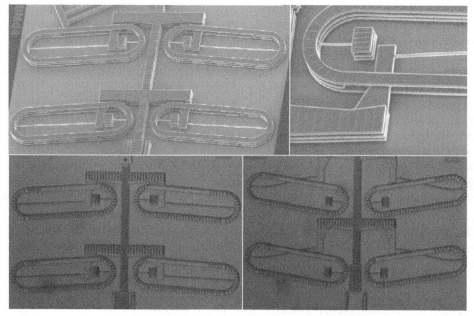

Figure 1.4 A multi-layer compliant microelectomechanical system (MEMS). A scanning electron micrograph of the device (top left) with a close up of compliant segments (top right), and the device shown in two stable equilibrium positions (bottom)

integration of different functions into fewer parts offers advantages, but it also requires the simultaneous design for motion and force behavior. This difficulty is increased further by the fact that the deflections are often well into the nonlinear range and simplified linear equations are not adequate to define their motion.

Fatigue life needs to be addressed for most compliant mechanisms. Because their motion comes from bending of flexible parts, compliant mechanisms experience stress at those locations. When that motion is repeated during its life, fatigue loads are present and the fatigue life must exceed the expected life of the mechanism. Fortunately, methods for analyzing and testing fatigue life are available to help design compliant mechanisms for their needed fatigue life (see Chapter 2), but it requires special attention and effort to ensure that the mechanism has the life required.

Although properly designed and tested compliant mechanisms can achieve needed fatigue life, there can still be a consumer perception that flexible components are flimsy or weak. This can be a particular concern where the flexible component is visible to the consumer and it may require special care in the design for adequate life and for its appearance.

The motion of compliant mechanisms is often more limited than for traditional rigid-link mechanisms. For example, a shaft connected to bearings has the ability to undergo continuous revolution, whereas the motion of a flexible component will be limited by the deflection it can undergo before failure.

The fact that strain energy is stored in a deflected beam can be either an advantage or a disadvantage. Advantages include that a compliant element integrates both a spring and hinge function into a single component providing a "home" position where the device will go when unloaded. This integration also allows certain behaviors, such as bistability (the characteristic of having two distinct preferred positions, such as the on-off positions of a light switch) [9]. However, there are times when these qualities are not desired, and the properties become a disadvantage in the device design.

If certain materials are held under stress for long periods of time or at elevated temperatures, they can take on a new shape associated with the stressed position. This is called "stress relaxation." Some compliant mechanisms have functions where they must maintain positions where they are under stress, and so are subject to stress relaxation conditions. This requires careful design and thoughtful choice of material.

1.4 Why are Compliant Mechanisms Becoming More Common?

Advances in our understanding of compliant mechanisms, combined with general technological developments, have resulted in a rapid growth in compliant mechanism applications (the library portion of this handbook is a testament to that growth). These applications range from high-end, high-precision devices to ultralow-cost packaging; from nanoscale featured components to large-scale machines; from weapons to healthcare products.

We mentioned that many early devices were compliant mechanisms, but then rigid-link devices connected at hinges gained favor because of the simplicity offered for analysis and design. So what is different now and why are there so many more compliant mechanisms than before? The answer lies at least partly in technological advances that have been made over recent decades. For example, new materials are available that are well suited for compliant mechanisms. There have been dramatic improvements in computational hardware and software available to analyze compliant mechanism motion and stresses. Developers and researchers have also increased our ability to design and analyze compliant mechanisms. Considerable effort has gone into creating methods to facilitate compliant mechanism design (some of the resulting methods are summarized in this handbook). There is also an increased awareness of the advantages of compliant mechanisms. As some commercial applications have been successful, they provide examples and inspiration for other applications to follow. Finally, as society and technology have advanced, new needs have risen, and some of these needs are best addressed by compliant mechanisms. This includes devices at very small size scales, devices with relatively complex motion but must be made at extremely low cost, compact medical implants, and high-precision machines.

1.5 What are the Fundamental Concepts that Help Us Understand Compliance?

There are a few straightforward but counterintuitive concepts that can help us understand the fundamentals of compliant mechanisms.

1.5.1 Stiffness and Strength are NOT the Same Thing

Usually when we want something to be strong (meaning that we don't want it to break), we also want it to be stiff (meaning that we don't want it to bend). For example, the floor in the upper story of a building we want to be both stiff and strong. We obviously don't want it to break, but we also don't want it to move around when people walk on it. So it needs to be stiff and strong. The crank shaft in an engine? Stiff and strong. A bridge? Stiff and strong. A desk? Stiff and strong.

   We so often design things that need to be both stiff and strong that it is easy for our intuition to begin to tell us that stiffness and strength are the same. But they are NOT the same. Strength relates to resistance to failure, while stiffness relates to resistance to deflection. These are different and are governed by different properties. Consider a piece of steel with a rectangular cross section as shown in Figure 1.5. The steel will withstand a certain stress until it will fail. But its strength is the same whether it is loaded about its thin or thick axis (assuming it is isotropic), while its stiffness is very different for these two conditions.

1.5.2 It is Possible for Something to be Flexible AND Strong

Consider examples of things that are both flexible and strong. Flexible endoscopes, such as that shown in Figure 1.6, are used to examine the interior parts of the body. The endoscope must be flexible to undergo the required motion and to minimize any trauma from its use within the body. It must also be strong to withstand the loads that it will undergo during its use. As another example, consider the pulleys on the cables of a ski lift (Figure 1.7). They must be strong enough to reliably lift the skiers to their destination but must be flexible enough to go around the pulleys.

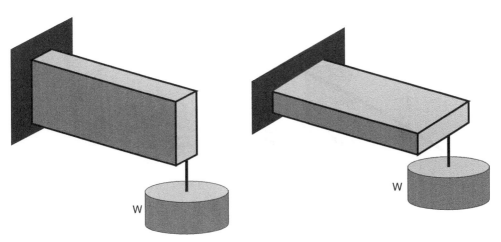

Figure 1.5 The rectangular piece of steel may have the same strength in different directions, but it will have very different stiffness for the two orientations shown

So why is it that many things we want to be stiff and strong, but others we want to be flexible and strong? What is it that determines the difference between these two situations? The answer lies in whether the device needs to hold a force, or if it needs to be deflected (like a cable going around a pulley). A bridge is an example of something that needs to be stiff and strong because we want it to hold the weight of

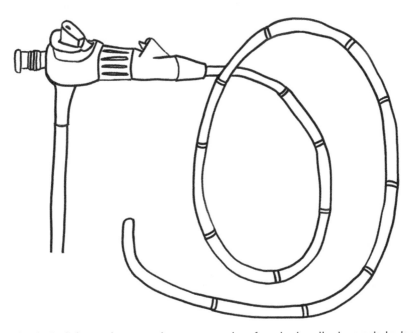

Figure 1.6 A flexible endoscope is an example of a device that needs to be both flexible and strong

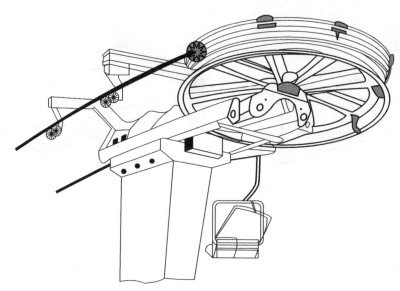

Figure 1.7 A ski lift cable must be flexible enough to go around the pulley and strong enough to carry the loads

traffic going across them without moving. The endoscope and the pulley cable are both examples of things that need to bend to perform their function. If they were too stiff, they would be overstressed and would break when they were forced to undergo the needed motion. So if something needs to hold a weight or other force, it should be stiff and strong; if it needs to go through a certain deflection, it should be flexible and strong.

1.5.3 The Basics of Creating Flexibility

There are three primary ways that we can influence flexibility. These are

1. material properties (what it is made of);
2. geometry (its shape and size);
3. loading and boundary conditions (how is it held and loaded).

Each of these is described below.

1.5.3.1 Materials Properties

Different materials have different stiffnesses as measured by the material's Young's modulus (or modulus of elasticity). Consider the three rods in Figure 1.8. Each rod has identical size and shape and each has the same size weight hanging from it, but they are each made of a different material: steel, aluminum, and polypropylene. The Young's modulus of steel (207 GPa) is about three times that of aluminum (72 GPa), so for the same geometry and same weight, the aluminum rod will deflect three times

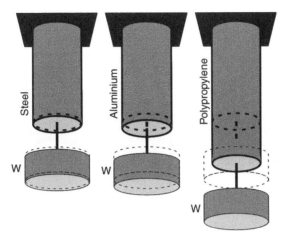

Figure 1.8 Material properties contribute to stiffness. These three rods all have the same geometry but they experience different deflections even with the same size weight hanging from them

as much as the steel. But polypropylene's Young's modulus (1.4 GPa), is about one fiftieth that of aluminum, so it deflects much farther than either aluminum or steel for the same applied load.

Because we want something both flexible and strong when we are designing compliant mechanisms, we look for materials with a high strength and a low Young's modulus. One way to compare different materials for use in a compliant mechanism is to compare the ratio of strength to Young's modulus, with a higher ratio being better. A similar approach is to compare the "resilience" of materials, where the modulus of resilience is one-half the yield strength squared divided by the Young's modulus. The modulus of resilience is a measure of how much energy per unit volume the material can withstand without a permanent change. Table 1.1 lists some materials and their ratio of strength to modulus and their resilience.

1.5.3.2 Geometry

Shape and size have a large effect on flexibility. Let's start with an obvious example. Consider the two parts shown in Figure 1.9. They are both made of the same material and they both have the same size of weight hanging from them. They are both round but one has a larger diameter than the other, and not surprisingly, the larger part is stiffer and has a smaller deflection than the smaller part. Now reconsider the two rectangular parts shown in Figure 1.5. They are made from the same material and have the same weight hanging from them. They are both the same size, but one is turned a different direction than the other. Even though they have the same size, the orientation of the geometry makes a huge difference in flexibility, with the part loaded on the thinner side being more flexible. Suppose the part were three times as wide as it were thick, then it would actually be nine times more flexible in one direction

Table 1.1 Yield strength to Young's modulus ratio and resilience for several materials

Material	E (GPa)	S_y (MPa)	$(S_y/E) \times$ 1000	$(0.5 \times S_y^2/E) \times$ 0.001
Steel (1010 hot rolled)	207	179	0.87	77
Steel (4140 Q&T @400)	207	1641	7.9	6500
Aluminum (110 annealed)	71.7	34	0.48	8.1
Aluminum (7075 heat treated)	71.7	503	7.0	1800
Titanium (Ti-35A annealed)	114	207	1.8	190
Titanium (Ti-13 heat treated)	114	1170	10	6000
Nitinol (high-temperature phase)	75	560	7.5	2100
Beryllium copper (CA170)	128	1170	9.2	5300
Polycrystalline silicon	169	930	5.5	2600
Polyethylene (HDPE)	1.4	28	20	280
Nylon (type 66)	2.8	55	20	540
Polypropylene	1.4	34	25	410
Kevlar (82 vol%) in epoxy	86	1517	18	13 000
E-glass (73.3 vol%) in epoxy	56	1640	29	24 000

than the other direction. If the geometry were distributed in other ways, such as in an I-beam shape, then it can be even stiffer for the same volume.

1.5.3.3 Loading and Boundary Conditions

Consider the three parts shown in Figure 1.10. Each part is made from the same material, has the same geometry, and the same size weight is hung from each. But the three parts will deflect differently for the same size weight. How the load is applied, and how the part is held (its boundary conditions), make a difference on its flexibility.

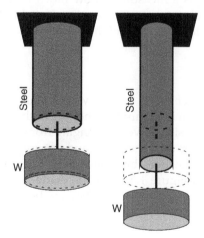

Figure 1.9 These two rods are both made of steel but their different geometries result in their having different deflections even for the same weight hanging from their ends

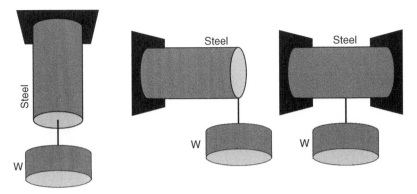

Figure 1.10 Boundary conditions and direction of loading also affect stiffness. The same rod deflects less in tension than in bending

1.6 Conclusion

Compliant mechanisms offer compelling characteristics that make them valuable for implementing in product and system design. Increased performance (e.g. high precision, low weight, compactness, low friction), reduced cost (e.g. reduced assembly, manufacturability), and ability to be miniaturized represent a few of the advantages of compliant mechanisms. The intent of this handbook is to provide information and tools that will be helpful in taking advantage of the possibilities of compliant mechanisms while addressing the challenges of their design.

References

[1] Howell, L.L., *Compliant Mechanisms*, John Wiley & Sons, New York, NY, 2001.
[2] Vogel, S., *Cats' Paws and Catapults*, W.W. Norton & Company, New York, NY, 1998.
[3] Taddei, M. and Zannon, E., Text by Domenico Laurenza, "Leonardo's Machines: Secrets and Inventions in the Da Vinci Codices", Giunti Industrie Graiche S.p.A, Firenze, Italia 2005.
[4] Wright, O. and Wright, W., "Flying Machine," U.S. Patent No. 821,393, May 22, 1906.
[5] Smith, S.T., *Flexures*, Taylor & Francis, London, UK, 2000.
[6] Lobontiu, N., *Compliant Mechanisms: Design of Flexures*, CRC Press, New York, NY, 2003.
[7] Wilcox, D.L. and Howell, L.L., "Fully Compliant Tensural Bistable Micro-mechanisms (FTBM)," *Journal of Microelectromechanical Systems*, Vol. 14, No. 6, pp. 1223–1235, 2005.
[8] Wilcox, D.L. and Howell, L.L., "Double-Tensural Bistable Mechanisms (DTBM) with On-Chip Actuation and Spring-like Post-bistable Behavior," *Proceedings of the 2005 ASME Mechanisms and Robotics Conference*, DETC2005-84697, 2005.
[9] Jensen, B.D., Howell, L.L., and Salmon, L.G., "Design of Two-Link, In-Plane, Bistable Compliant Micro-Mechanisms," *Journal of Mechanical Design*, Trans. ASME, Vol. 121, No. 3, pp. 416–423, 1999.

2

Using the Handbook to Design Devices

Spencer P. Magleby

Brigham Young University, USA

As described in Chapter 1, compliant mechanisms (CMs) can provide a variety of aesthetic, functional, fabrication and maintenance advantages over traditional rigid-body mechanisms in many situations. Chapter 1 also makes it clear that there are some special concerns and trade-offs to consider when designing compliant mechanisms as compared to more traditional mechanisms. We will discuss a few of those in this chapter and how to minimize their impact. There are also challenges in synthesizing and modeling compliant mechanisms as forces and deflections are coupled. In the chapters that follow we will provide tools and approaches to address these challenges, and later in this chapter show some basic design decision processes that can be used to choose among the various tools. Finally, the handbook library is a vast source of inspiration, basic mechanism elements and starting points for creating your own compliant mechanisms.

The handbook has been fashioned for the primary audience of practicing engineers that have some knowledge of designing mechanisms for specific applications. We have also worked to assure that it has value for designers and inventors that are not engineers, and that they can gain inspiration from the handbook – through examples and illustrations in the chapters and especially through the entries in the library. For engineering designers that need to meet specific performance requirements, we assume that they have access to texts on compliant mechanisms (some are recommended at the end of this chapter). Many of the chapters in the handbook have a section of recommended readings that can enhance and expand on the information presented by the authors.

The overall objective of the handbook is to inspire confidence in the reader/user that they can successfully develop compliant mechanisms for specific applications. The focus in the chapters to follow is on practice and practicality, with enough background to allow engineers and designers to apply the examples in the library and perhaps extend the examples to more complex situations and systems.

2.1 Handbook Outline

The handbook is divided into four main Parts. The first three Parts are intended to orient the user of the handbook and provide general information on the modeling and synthesis of compliant mechanisms. The fourth Part constitutes the bulk of the handbook and contains a structured library of compliant mechanism elements, devices and systems. The objectives of the Parts are summarized below:

I. *Introduction to Compliant Mechanisms* – The two chapters in this Part (including this chapter) introduce the reader to compliant mechanisms, provide motivation for their use and then show how to employ the handbook to select, synthesize, model and design compliant mechanism elements and devices to meet specific needs.

II. *Modeling of Compliant Mechanisms* – This Part focuses on modeling the behavior and performance of complaint elements and mechanisms. The Part is broken into three chapters. The first two chapters address closed-form modeling of elements with small/intermediate deflections and large deflections. The third chapter addresses a modeling technique that allows compliant mechanisms to be modeled using rigid-body modeling techniques and can be applied both to individual elements and more complex mechanisms. Modeling is emphasized in the handbook as it is the starting point for adapting concepts found in the library to a given situation, or to start "from scratch" on a compliant mechanism based on inspiration found in the library.

III. *Synthesis of Compliant Mechanisms* – In this Part four synthesis approaches are presented that show both how mechanisms and systems in the library were generated, and how an engineer/designer can synthesize variations of the library entries – or new mechanisms "from scratch". The first chapter in this Part focuses on synthesis of mechanisms that will undergo relatively small deflections, while the other are generally more suitable for larger deflections. These synthesis methods are reviewed below.

IV. *Library of Compliant Mechanisms* – This Part consists of a description of the library organization and the library itself.

2.2 Considerations in Designing Compliant Mechanisms

While there are clearly a number of significant potential advantages of compliant mechanisms over traditional rigid-body systems, as with any engineered system there are trade-offs to be made as concepts are selected and design parameters are chosen. Many of the concepts described in the library were developed to be optimal in some situation – that is to minimize the trade-offs and maximize the desired performance. Outlined below are three inter-related areas where designers commonly need to carefully consider trade-offs for compliant mechanisms: fatigue failure, achieving large deflections and maintaining off-axis stiffness. The section concludes with a brief discussion of a unique consideration in designing compliant mechanisms, the coupling of forces and deflections.

Fatigue Failure

Many designers considering the use of CMs are concerned about fatigue failure as we have generally been trained to avoid repeated deflections of materials – especially large deflections. While the fatigue life of a CM is of course a concern, there are many ways to mitigate fatigue and still achieve desired performance levels. Fatigue failure of the flexible elements of CMs can occur from tension, compression, torsion or bending, but bending, and sometimes torsion, are generally the dominant considerations. Here, we focus on bending, but torsion or combined stresses can be treated in a similar manner.

In bending, fatigue life will be closely correlated with the maximum stress which is a function of the deflection and moment of inertia of the beam. Since deflection is what the designer is trying to achieve (see thoughts below on ways to reduce required deflections), the means to lowering stress and increasing fatigue life is to reduce the appropriate moment of inertia. This is generally done by making the beam "thinner". As discussed in Chapter 1, designers must avoid the natural tendency to stiffen elements to reduce stress, and instead think in terms of reducing stress levels for a given deflection. It should be noted that as the moment of inertia is reduced the stiffness of the beam is also reduced, changing the response of the beam to loadings. This coupling effect is discussed below. In addition, lowering the stiffness about one axis may lower the levels of desired stiffness in other directions. Managing this ratio of desired and undesired stiffness is also reviewed below.

Lastly, designers should carefully select materials to reflect a balance of fatigue life, stress limits, deflection and other performance requirements. Designed correctly, CMs – even those with large deflections – can meet very demanding requirements for load/deflection cycles. For example, the light switch device design described in M-93 of the library was tested to over one million cycles without failure in our laboratory when fabricated from polypropylene. Many entries in the library were conceived of to specifically address high stress levels.

Achieving Large Deflections

For designers new to CMs it may be difficult to imagine that a device will be able to achieve predictable, large motions through deflections of its elements. However, even many everyday devices, such as shampoo cap lids (shown in library entry EM-17) exhibit large deflections. For a new designer it is helpful to think about three basic ways to achieve large deflections in common CM configurations:

1. Reducing the moment of inertia of an element in bending (or polar moment of inertia for torsion) is the most straightforward way to achieve larger deflections. This approach is highly related to the fatigue discussion above and the coupling discussion below. While this approach may seem obvious, *reducing* the moment of inertia can be nonintuitive and is often overlooked.

2. Increasing the length of the element in bending or torsion will increase the deflection for a given loading without increasing stress levels (and hurting fatigue life). While this approach will increase deflections, it will almost always decrease the off-axis stiffness as discussed below. Nevertheless, the maximum deflection possible for many CM designs is highly influenced by the length of deflecting members.
3. If individual members cannot practically achieve the desired deflections, the designer may choose to arrange deflecting elements in series, thus requiring less deflection from an individual member. An example is entry M-56 in the library. The elements in series do not have to be the same – or even use the same deflection mode. A good example is the Lamina-Emergent Torsional Joint shown in library entry EM-35. This joint employs both bending and torsion in series to achieve larger deflections. It is also interesting to note the use of parallel elements to improve the off-axis stiffness of the joint as is mentioned in the section below.

The handbook contains other means to achieve large deflections that are more sophisticated or involve more complex devices. In many cases the three approaches above are used in various combinations to achieve the desired performance. Recognizing these fundamental strategies will help designers adapt them to their own needs.

Maintaining Off-Axis Stiffness

A common assumption in the synthesis and modeling of rigid-body mechanisms is that all motion will occur at the joints and that links are infinitely rigid. This assumption generally allows us to use *kinematics* to describe and model the motion behavior of a rigid-body mechanism. For CMs, motion occurs because elements of the system are allowed to deflect under load to achieve desired behaviors. Any localization of the deflection (such as within a specific mechanism) is due to the lower stiffness of the deflecting element relative to the stiffness of other members – or of the same member in other directions.

Let's consider a compliant element or system that has been designed to allow rotational motion about one axis. We define the off-axis stiffness ratio as the ratio of the stiffness about an undesirable axis relative to the stiffness about the desirable axis of motion. If the stiffness ratio is high then the localization is significant. If the ratio is low then the localization is less prominent, meaning that the element is more prone to move in undesirable directions under loading. In many applications maintaining a high off-axis stiffness ratio is a key design objective. As an example consider a hinge for an access door. The rigid-body version of the hinge-door system localizes all rotational motion about the axis of the hinge pin. The system is relatively stiff in any other direction and for many practical purposes could be considered rigid in all directions except the rotation required to open the door. If a CM was used to allow the door to open, it would likely be designed to have relatively low stiffness about the rotational axis required to open the door – perhaps using a series of flexible segment. This type of CM hinge will likely be prone to some translation along the hinge rotation axis and rotation about an axis perpendicular to the hinge axis. The ratios of these stiffnesses will have a large effect on the performance of the system. Note that the ratio can be influenced both by reducing the stiffness in the axis of desired rotation and increasing the other stiffnesses.

Many CM element and device designs in the library have been specifically created to help assure high off-axis stiffness ratios. Consider, for example, library entries EM-8 and EM-13. These mechanisms contain elements that control the deflection about the desired axis, as well as elements that help assure high levels of stiffness in undesirable directions.

Coupling of Forces and Deflections

When comparing the design of CMs with rigid-body mechanisms, it is essential to note the inherent coupling of forces and movement. In the discussion above we mention that when considering rigid-body mechanisms, the designer can usually separate motion (the kinematics) from the transmitted forces (the kinetics). This is generally not possible for CMs as forces are required to produce deformations (motion) and the deformations are dependent on the materials and geometry of the mechanism. Thus, the designer usually cannot independently design the kinematics of a CM and the force–deflection relationships. The modeling and synthesis techniques and approaches in the handbook chapters show a variety of means for handling this coupling.

While this relationship may complicate the synthesis and analysis of a CM, it can also create the potential for unique and efficient performance characteristics in the mechanisms. A good example is the bistable characteristic of the switch shown in library entry M-93. In this case the energy stored in the deflecting members of the mechanism causes the switch to "pop" into two different positions without the use of a spring that is external to the mechanism. With proper design, CMs can be inherently biased into particular configurations, absorb energy and release energy.

2.3 Locating Ideas and Concepts in the Library

The handbook library contains hundreds of entries that can be used as a source of basic compliant elements or entire mechanisms. In general, the intent is to provide the designer with ideas and inspiration that can support design processes, and/or be the starting point for designing new mechanisms. The basic compliant elements, such as flexible beams, can be used as building blocks for developing CMs, or as elements that can substitute for rigid-body elements (see Chapters 8 and 9).

The introductory materials to the library section outline the organization of the library entries and the classification scheme that was used to index the entries. Designers can then search for mechanisms in a variety of ways depending on their level of understanding of the mechanism function they are considering. If you are seeking inspiration in general then it may be best to look over those entries that are complete devices (such as a tennis racquet) or are oriented to specific applications. If the functions of the CM being designed are known then it may be efficient to search focused mechanisms or joints by degrees of freedom or type of motion. If the mechanism requires specific behaviors, such as bistability, as mentioned above, then it may be best to search for complete mechanisms that can be used for starting points.

In most cases the designer will note that there are a variety of ways shown in the library for achieving a given motion or force–deflection relationship. The various means reflect the desire to favor different trade-offs as discussed above. In addition to function, selection may also be guided by manufacturability, material constraints and operating conditions.

2.4 Modeling Compliant Mechanisms

Once a concept is chosen it must be modeled in such a way as to help the designer choose acceptable values for design parameters (such as thickness and length of a flexible beam) or confirm the performance of a given design. The modeling approaches discussed in Part II of the handbook were selected because they are focused on techniques that are conducive to design of CMs through rapid iteration of analysis of design performance (at elemental or mechanism levels) and/or determination of key performance relationships and trends. The modeling approaches can be better understood through classification by amount of relative deflection, and the use of closed-form vs. approximation techniques.

Selection Based on Deflection Levels

The expected amount of deflection of a CM element relative to its key dimensions can help to determine the most appropriate modeling approach. In the handbook chapters on modeling we define small deflections to be significantly less than 10% of the beam length, intermediate deflections to be in the range of 10% and large deflections to be those exceeding 10% of beam length. Of course, there is significant overlap in these categorizations as the modeling tools have advantages and disadvantages that make them attractive through a range of displacements.

- For small deflections the designer may be able to use traditional deflection analysis with the small-angle assumptions for beam bending. Generally, configurations with very small deflections would be considered structures, not mechanisms, and as such do not receive significant attention in the handbook.
- For intermediate deflections the designer will want to reference the material in Chapter 3. Devices with deflections in this range are often considered precision devices and are common in positioning and measurement systems. For these applications accuracy is often highly important.
- For larger deflections the approach outlined in Chapter 5 is likely the easiest to use and most similar to traditional rigid-body mechanism modeling. Modeling of devices with deflections in this range generally favors flexibility and intuition over precision.

Closed Form versus Approximate Modeling

Some applications of CMs, especially those implemented for precision equipment, will benefit from the closed-form modeling approaches described in Chapters 3 and 4. These techniques are generally the most accurate, but likely most applicable for smaller and precision movements. They are also best used for analyzing individual elements as the modeling can quickly become complex for larger mechanism systems. For larger deflections and more complex mechanisms systems the designer may choose to use the pseudo-rigid-body model approach described in Chapter 5. This approach models CM behavior using a simplifying assumption (thus introducing some small approximation), but retains sufficient accuracy for most situations.

An important modeling tool that is not discussed in the handbook is finite element analysis (FEA). A wide array of literature already exists for FEA so it is not reviewed in the handbook. Properly applied, FEA modeling can be an excellent way to check the predictions of one of the modeling techniques discussed above, or fine tune a design for optimal performance. Designers should be aware that the relatively large deflections common with CMs require special considerations when choosing modeling elements, applying loads and setting boundary constraints.

2.5 Synthesizing Your Own Compliant Mechanisms

The handbook library contains a large number of CM elements and devices, but this is just a sampling of the many available and yet-to-be developed examples of CMs. Chapters 6, 7, 8 and 9 of the handbook have been included to illustrate techniques and approaches to synthesizing new compliant mechanisms most appropriate for specific applications or as part of larger devices. The three sections below discuss synthesis techniques from the simplest to the more complex:

Modifying a Concept Found in the Library
Perhaps the simplest way to synthesize a CM is to merely modify a concept found in the handbook library or combine concepts. This is generally most effective if the concept being modified has the needed function(s), but the designer wants a different set of trade-offs in flexibility, deflection, etc. Basic elements of the library concept (such as particular joints) can be replaced with alternative elements found in the first sections of the library.

Replacing an Existing Rigid-Body Mechanism
In many situations the designer is considering replacing an existing rigid-body mechanism design or hardware that is performing a function. In these cases a quick way to a viable design is likely through the use of rigid-body replacement synthesis. Chapter 8 illustrates this technique and provides an example. Rigid-body replacement is especially attractive as a way to quickly explore a variety of configurations. In addition, it potentially takes advantage of the experience and background of a traditional mechanism designer.

Starting from Functional Requirements
In many cases the designer may choose to synthesize a CM starting with the basic functional requirements for the mechanism. This approach has the advantage of creating new ideas and potentially achieving a more optimal match of the CM and

the requirements. Three techniques are introduced in the chapters – each with its own strengths and application area.

Chapter 6 presents a structured technique called freedom and constraint topologies for synthesizing CMs with small to intermediate deflections when functional needs, especially desired degrees-of-freedom, are well known. The technique builds on its own library of geometric shapes that can be combined to realize basic mechanisms that exhibit specific degrees-of-freedom while preserving off-axis stiffness. Designers considering development of new mechanisms for precision devices will want to first consider this technique.

Chapter 9 also uses the idea of building blocks to synthesize CMs, but with application that might be better suited for mechanisms with intermediate to large deflections. The building blocks in this synthesis approach are linked to basic functions of a mechanism and can be combined in various ways to achieve the goals of the designer. The intuitive nature of the building blocks will be attractive to designers accustomed to working with traditional mechanisms.

The most general of the synthesis techniques in the handbook, topology optimization, is outlined in Chapter 7. This technique requires that the designer have a clear idea of the desired inputs and outputs of the CM being designed. From this information computer algorithms are then employed to search for the shape (topologies) that will best achieve the desired results. Topology optimization can yield unique CM configurations that a designer would be unlikely to develop by combining known elements together. An example of a CM developed using shape optimization can be seen in entry M-45 of the library. This technique will be especially attractive to designers that have specific functional needs in mind and are willing to consider unusual shapes and geometries.

While the techniques outlined in Chapters 6–9 are not the only ways to synthesize CMs, they do represent structured approaches that are likely to result in desired behaviors. Designers may choose to combine these approaches with other general mechanical synthesis techniques to develop solutions to motion and force requirements for specific applications.

2.6 Summary of Design Approaches for Compliant Mechanisms

As a summary of the discussions above, basic decision processes for selecting concepts and determining design parameter values are outlined next.

Selecting a Concept

Selecting or synthesizing a concept for a CM can appear a bit daunting as there are a large number of mechanisms possible for a given kinematic and force/deflection need. In most situations the designer new to CMs should seek a practical and sufficient concept instead of an optimal solution. More experienced designers may choose more sophisticated approaches.

As a summary, three cases of starting points for concept selection are listed below. Each case description is followed by recommended actions for developing a CM solution.

Case 1: There is an existing rigid-body mechanism performing a desired function(s).
            The first approach would be to pursue rigid-body replacement (Chapter 8) using the rigid-body mechanism as a starting point. If this approach does not yield desirable results then the designer might look through complete mechanism concepts in the library that have motions similar to the rigid-body mechanism.
Case 2: A function has been identified that requires a mechanism, but no rigid-body mechanism has been designed.
            This is a chance to find a mechanism in the library that roughly achieves the desired functions. From this point rigid-body replacement could again be used to make modifications. If the functions require high-precision, low-deflection movements then it may be effective to go directly to the basic synthesis approach outlined in Chapter 6. If the functional needs are complex then an appropriate synthesis technique should be chosen using the guidance in Section 2.5 above.
Case 3: A compliant mechanism exists, but is not performing acceptably.
            Before moving to a completely new CM it could be effective to look for elements in the library, especially joints, which could be substituted for existing elements. If the designer feels that a wholly new idea is called for, then starting with new building blocks or using topology optimization may be appropriate.

Of course, you may also be interested in perusing the library just to get a feel for what might be possible with compliant mechanisms – even if you do not have a specific application in mind or know the exact specifications that you are looking for.

Determining Design Parameter Values

Once a concept has been selected, the designer must move to determining appropriate parameter values – especially those that have a large influence on the performance of the CM. Given the coupling of deflection and forces discussed above, it is advisable to first choose a material, then select a modeling approach and finally use the model to determine acceptable values.

Considerations in the selection of materials for a CM are more completely discussed in the text by Howell listed in the readings below. Material properties will have a very large effect on the behavior of a CM, especially in situations where large deflections are desired. Balancing stress limits, modulus and fatigue life are the key functional considerations.

With the material chosen, the designer must then select a modeling approach. Section 2.4 above discusses the basics of this decision. If the designer feels that the design space to be explored is large then modeling approaches that can show trends and support designer intuition should be chosen.

With the model in place the designer can now explore the behavior of the CM with different parameter values. For all but the simplest of compliant elements it is generally impractical to compute a single solution to achieving a given behavior. For typical CMs it then becomes necessary to explore the possible design space for the key parameters. For this to be effective, the number of parameter values to be considered should be reduced by decomposing the system, making dimensionless ratios of values (such as thickness of a beam divided by length of the beam), or assuming fixed values of some parameters. Spreadsheets and basic optimization approaches are often useful at this stage.

Once a set of acceptable parameter values has been determined, many designers choose to make a final performance check using Finite Element Analysis (FEA). In some cases, limited design iteration can be done at this stage using FEA. In our lab we also like to check models and designs using prototypes. It is often possible to substitute material properties into the model that correspond to materials that are amenable to the fabrication of prototypes (such as polypropylene), and then checking the physical behavior of the prototype against the model predictions.

Further Reading

A variety of texts and academic papers are available for further reference on broad aspects of synthesizing, modeling and designing compliant mechanisms. These materials can be complimentary to the information contained in the handbook chapters. Below are listed some pertinent materials developed by researchers from the Compliant Mechanisms Research Group at Brigham Young University. The authors of the following chapters have also provided citations of supporting materials for their topic areas. These materials can be a starting point for seeking out information from researchers and developers throughout the world.

There are a few textbooks available on compliant mechanisms that develop the foundations of compliant mechanisms. These generally have more details on modeling and analysis of CMs and could be a valuable companion to this handbook for those that would like more background or details on the techniques overviewed in chapters. Three widely cited books are listed below:

L.L. Howell, *Compliant Mechanisms*, John Wiley & Sons, Inc., New York, NY, 2001.
N. Lobontiu, *Compliant Mechanisms: Design of Flexure Hinges*, CRC Press, New York, NY, 2002.
S.T. Smith, *Flexures: Elements of Elastic Mechanisms*, Taylor and Francis, London, UK, 2003.

Within our research group we have published materials that describe in detail design process or tools that could be used for compliant mechanisms. Selected references are listed below that can help the reader get a start on understanding this area and search for work by other groups.

Mattson, C., Howell, L. and Magleby, S., "Development of commercially viable compliant mechanisms using the pseudo-rigid-body model: Case studies of parallel mechanisms," *Journal of Intelligent Materials Systems and Structures*, March 2004, Vol. 15, no. 3, pp. 195–202(8).
Berglund, M., Magleby, S. and Howell, L., "Design Rules for Selecting and Designing Compliant Mechanisms for Rigid-Body Replacement Synthesis," Proceedings of the 26th ASME Design Engineering Technical Conferences, Design Automation Division, Baltimore, MD, September, 2000, DETC2000/DAC-14225.

Mackay, A., Smith, D., Magleby, S., Howell, L. and Jensen, B., "Metrics for evaluation and design of large-displacement linear-motion compliant mechanisms," *ASME Journal of Mechanical Design*, Vol. 134, Issue 1, 011008 (9 pages), 2012.
Ferrell, D., Isaac, Y., Magleby, S., and Howell, L., "Development of criteria for lamina emergent mechanism flexures with specific application to metals," *ASME Journal of Mechanical Design*, Vol. 133, Issue 3, 031009 (9 pages), 2011.
Olsen, B., Howell, L., Magleby, S, "Compliant Mechanism Road Bicycle Brake: A Rigid-body Replacement Case Study," *Proceedings of the ASME International Design Engineering Technical Conferences*, Washington, DC, August 28-31 2011, DETC2011-48621.

Along with other researchers, our research group has published materials on applications of compliant mechanisms to specific product areas. A sampling are listed below that can help the interested reader get started and search out a range of applications and researchers.

Crane, N.B., Howell, L.L., Weight, B.L. and Magleby, S.P., "Compliant floating-opposing-arm (FOA) centrifugal clutch," *Journal of Mechanical Design, Trans. ASME.*, Vol. 126, No. 1, pp. 169–177, 2004.
Fowler, R., Howell, L.L. and Magleby, S.P., "Compliant space mechanisms: a new frontier for compliant mechanisms," *Journal of Mechanical Sciences*, 2, pp. 205–215, doi:10.5194/ms-2-217-2011, 2011.
Guerinot, A.E., Magleby, S.P., Howell, L.L., Todd, R.H., "Compliant joint design principles for high compressive load situations," *Journal of Mechanical Design*, Vol. 127, No. 4, pp. 774–781, 2005.
Guerinot, A., Magleby, S. and Howell, L., "Compliant Mechanisms Concepts for Prosthetic Knee Joints," Proceedings of the 2004 ASME International Design Engineering Technical Conferences, Mechanisms and Robotics Conference, Symposium on Medical Devices and Systems, Salt Lake City, Utah, September 2004, DETC2004-57416.
Weight, B., Mattson, C., Magleby, S. and Howell, L., "Configuration selection, modeling, and preliminary testing in support of constant-force electrical connectors," *Journal of Electronics Packaging, Transactions of the ASME*, Vol. 129, September 2007, pp. 236–246.

Part Two
Modeling of Compliant Mechanisms

3

Analysis of Flexure Mechanisms in the Intermediate Displacement Range

Shorya Awtar

University of Michigan, USA

3.1 Introduction

This chapter presents a nonlinear, parametric, closed-form model for a planar (or 2D) beam that is accurate over an intermediate range of displacements, typically 10% of the beam length. This nonlinear beam model, referred to as the beam constraint model, enables deterministic analysis and optimization of flexure mechanisms (another equivalent term for compliant mechanisms), helps identify their performance limits and tradeoffs, and better informs their constraint-based synthesis.

Flexure mechanisms provide guided motion via elastic deformation and are used in a variety of applications that demand high precision, minimal assembly, long operating life, and/or design simplicity. The motion guidance functionality of a flexure mechanism results in degrees of freedom (DoF) and degrees of constraint (DoC), analogous to those seen in traditional rigid-link mechanisms. In the case of flexure mechanisms, the DoF directions are associated with small stiffness while the DoC directions exhibit several orders of magnitude higher stiffness.

As an example, the parallelogram flexure mechanism (from L.L. Howell, *Compliant Mechanisms*, John Wiley & Sons, Inc., 2001) is shown in Figure 3.1A. Here, two parallel flexure strips are configured between a fixed ground and a motion stage such

Handbook of Compliant Mechanisms, First Edition. Edited by Larry L. Howell, Spencer P. Magleby and Brian M. Olsen.
© 2013 John Wiley & Sons, Ltd. Published 2013 by John Wiley & Sons, Ltd.

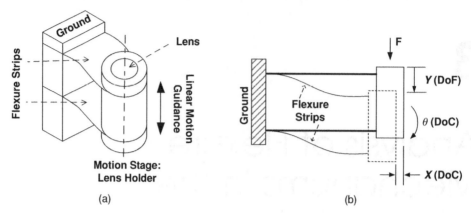

Figure 3.1   Parallelogram flexure mechanism

that the latter has a single translational DoF along the Y-direction. All other directions are constrained and therefore represent DoC. This frictionless and backlash-free flexure mechanism guides an objective lens mounted on the motion stage along the Y-direction, following an approximately linear trajectory, to allow precise focus adjustments in an optical assembly. A planar (or 2D) stick-figure representation of this flexure mechanism is shown in Figure 3.1B.

For a given flexure mechanism, such as the one shown above, a designer often seeks to determine its displacement range prior to material failure, stiffness along the DoF and DoC directions, variation in these stiffness values with increasing loads and displacements, and undesired or error motions along the DoC directions. The designer may also seek to understand how these motion guidance attributes are dependent on the geometric parameters (i.e. dimensions) of the flexure mechanism that would allow design optimization. These goals motivate the need for predictive, analytical modeling of flexure mechanisms.

However, before delving further into modeling, there are certain observations that can be made qualitatively. For example, in the flexure mechanism of Figure 3.1:

1. There exist error motions along the X and Θ DoC that increase with increasing displacement along the Y DoF.
2. A tensile force along the X DoC increases the stiffness along the Y DoF and a compressive force does the opposite.
3. The X and Θ DoC direction stiffness drops with increasing displacement along the Y DoF.

Extensive analytical and experimental results have shown that the above attributes, which directly affect the motion guidance performance of the flexure mechanism, are strongly dependent on geometric nonlinearities in flexure mechanics. Therefore, while a linear elastic load–displacement model is simple to derive, closed-form, and parametric, it fails to capture these observations. Inclusion of geometric nonlinearities in the analytical modeling of flexure mechanisms is generally nontrivial. Although

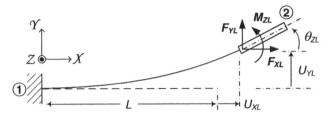

Figure 3.2 Beam flexure

numerical methods such as nonlinear finite element analysis (FEA) may be used to obtain accurate results, they offer little parametric design insight. Instead, an analytical model is desirable that is simple, closed-form, and parametric on the one hand, but also accurately captures the relevant geometric nonlinearities for general loading conditions over a practically useful range of displacements.

In principle, if the basic constituent elements of a flexure mechanism are modeled, then the entire flexure mechanism can also be modeled via appropriate mathematical steps. This then shifts the attention to the modeling of constituent elements, of which the most prominent one is the beam flexure (Figure 3.2). Given its long and slender geometry, a beam flexure offers low stiffness in the transverse bending direction and high stiffness in the axial stretching direction, and therefore serves as a useful constraint element or building-block in various flexure mechanisms.

3.2 Modeling Geometric Nonlinearities in Beam Flexures

The standard mechanics formulation for long, slender, planar beams is based on the Euler and Bernoulli assumption that *"plane cross-sections that are perpendicular to the beam centroidal axis prior to deformation remain plane and perpendicular to the neutral axis after deformation"*. This assumption rules out shear deformations even in the presence of shear loads. A standard mechanics formulation based on this assumption leads to the following governing relation for the beam shown in Figure 3.2:

$$\frac{E}{\rho(X)} = \frac{M_Z(X)}{I_{ZZ}} \tag{3.1}$$

Here, E is the Young's modulus for beams with depths comparable to its inplane thickness and plate modulus for beams with depth several orders of magnitude greater than its inplane thickness. This relation applies to every cross section: M_Z is the moment, I_{ZZ} is the second moment of area about the Z-axis, ρ is the radius of curvature at a given cross section that was located at coordinate X prior to deformation. While nonlinearities can arise from material properties, the above expression assumes linear constitutive relations between stresses and strains. Such linear material behavior is a reasonable assumption for most engineering materials.

To employ the above relation in generating the load–displacement relations at the beam end, there are three steps needed:

- First, the bending moment at the relevant cross section has to be expressed in terms of the beam end-loads, which is done by applying the load equilibrium conditions. The load equilibrium may be applied in the undeformed configuration of the beam as follows:

$$M_Z(X) = M_{ZL} + F_{YL}(1 - X) \tag{3.2}$$

This is justified only when the displacements of the beam are infinitesimally small. The most accurate expression for load equilibrium is obtained when it is applied in the deformed configuration of the beam, as follows:

$$M_Z(X) = M_{ZL} + F_{YL}(1 + U_{XL} - X) - F_{XL}(U_{YL} - U_Y(X)) \tag{3.3}$$

Applying load equilibrium in the deformed beam configuration is mathematically and physically equivalent to recognizing the contribution of rotation of cross sections to axial strain or equivalently recognizing the beam arc-length conservation, all of which are important geometric nonlinearities.
- Secondly, the beam curvature $\rho(X)$ has to be expressed in terms of the position coordinates and displacement variable of the beam. For deformations within 10% of the beam length, the curvature may be approximated as:

$$\frac{1}{\rho(X)} = Y''(X) \tag{3.4}$$

However, the most mathematically accurate expression for curvature is given by:

$$\frac{1}{\rho(X)} = \frac{Y''(X)}{(1 - Y'(X)^2)^{+1/2}} \tag{3.5}$$

Clearly, this represents another important geometric nonlinearity.
- Finally, the appropriate load equilibrium and curvature expressions are substituted into Eq. (3.1) to produce a beam-governing differential equation, which is then solved in the presence of appropriate boundary conditions. This leads to the necessary end-load displacement relations.

In the mathematically simplest case, one can substitute the approximations of expression (3.2) for load equilibrium and expression (3.4) for curvature in Eq. (3.1) to obtain the following completely linearized beam governing differential equation:

$$EI_{ZZ}U_Y''(X) = M_{ZL} + F_{YL}(1 - X) \tag{3.6}$$

This is the beam-bending equation traditionally found in standard textbooks. Assuming uniform cross section along the beam length, i.e. I_{ZZ} independent of X, this equation can be solved using the boundary conditions at the fixed end of the beam $U_Y(0) = U'_Y(0) = 0$ to produce the following load–displacement relations at the free end of the beam:

$$
\begin{bmatrix} \dfrac{F_{YL}L^2}{EI_{ZZ}} \\[2ex] \dfrac{M_{ZL}L}{EI_{ZZ}} \end{bmatrix} = \begin{bmatrix} 12 & -6 \\ -6 & 4 \end{bmatrix} \begin{bmatrix} \dfrac{U_{YL}}{L} \\[2ex] \theta_{ZL} \end{bmatrix} \quad \text{where} \quad \theta_{ZL} = U'_{YL} \tag{3.7}
$$

Separately, one can apply Hooke's Law in the X-direction to produce the following

$$
\frac{F_{XL}}{EA} = \frac{U_{XL}}{L}
$$

$$
\Rightarrow \frac{F_{XL}L^2}{EI_{ZZ}} = \frac{12}{(T/L)^2} \frac{U_{XL}}{L} \tag{3.8}
$$

Together, the final results (3.7) and (3.8) do not capture any geometric nonlinearities in the individual beam flexure, and therefore do not help predict the motion guidance behavior of flexure mechanisms discussed previously. For the X-direction, result (3.8) is accurate only when the transverse displacements (U_{YL} and $L^*\theta_{ZL}$) are of the order of the thickness T of the beam. In the transverse or bending direction, the relations (3.7) are accurate only when the axial load F_{XL} is negligible and the transverse displacements (U_{YL} and $L^*\theta_{ZL}$) are within 10% of the beam length L.

Alternatively, one may employ the accurate nonlinear load equilibrium expression (3.3) and curvature expression (3.5). Substituting these into Eq. (3.1) produces the fully nonlinear beam governing differential equation:

$$
\frac{EI_{ZZ}Y''(X)}{(1-Y'(X)^2)^{+1/2}} = M_{ZL} + F_{YL}(1 + U_{XL} - X) - F_{XL}(U_{YL} - U_Y(X)) \tag{3.9}
$$

Solving this nonlinear equation, along with the previously stated boundary conditions, for general end-loads is mathematically nontrivial. For specific end-loads and uniform beam cross section, solution procedures based on elliptic integrals are discussed in considerable detail in Chapter 4. However, the final solutions for end-displacements in this approach have to be obtained numerically, making it too complex for flexure mechanism design. Displacement solutions for any general end-loads may also be obtained using a nonlinear finite elements analysis (FEA) that incorporates all of the above geometric nonlinearities. Both these nonlinear solution approaches, although very accurate, offer little parametric design insight.

This concern is addressed, in part, by the pseudo-rigid body model (PRBM), which is the subject of discussion in Chapter 5. The PRBM represents a lumped-parameter modeling approach to capture the large displacement behavior of beam flexures and is derived from an optimization process that utilizes the exact nonlinear solution for a beam flexure that might have been obtained via elliptic integrals or other numerical methods. For this reason, the PRBM parameters must be recomputed for every change in the loading and boundary conditions. Once obtained, the PRBM accurately captures the transverse direction load–displacement relations over a very large displacement range: U_{YL} and $L^*\theta_{ZL}$ of the order of the beam length L. Furthermore, it captures the stiffening of the DoF directions in the presence of DoC loads as well as the purely kinematic or geometric components of error motions in the DoC directions. However, the inherent lumped-compliance assumption of a PRBM precludes any stiffness variation in the DoC direction with increasing DoF displacements, and certain DoC direction error motions. These observations are quantitatively derived and graphically illustrated later in this chapter.

To overcome these limitations, the approach presented in this chapter is to carry out a partial linearization of the beam governing equation, i.e. employ the linearized approximate expression (3.4) for curvature but the nonlinear accurate expression (3.3) for load equilibrium. Since flexure mechanisms typically employ long slender beams that undergo DoF displacements that are within 10% of the respective beam lengths, the beam curvature nonlinearity is not of much significance (<1% approximation error). However, the presence of an axial force F_{XL} that is comparable to the transverse loads (F_{YL} and M_{ZL}/L) produces as large as a 10% contribution to the bending moment at a given cross section, and therefore may not be ignored. Also, as noted earlier, the geometric nonlinearity associated with load equilibrium is implicitly equivalent to the beam arc-length conservation nonlinearity, which is critical to capture the kinematics of beam-flexure deformation.

The outcome of this partial linearization is that one obtains a model, referred to as the beam constraint model (BCM), that on the one hand is simple, closed-form, parametric, and incorporates any general end-loads; on the other hand, it captures all the relevant geometric nonlinearities over an intermediate range of transverse displacements (U_{YL} and $L^*\theta_{ZL}$ within 10% of the beam length L). This beam constraint model is presented in the next section, and its effectiveness in predicting all the relevant motion guidance attributes of flexure mechanisms is demonstrated in Section 3.4.

3.3 Beam Constraint Model

Substituting the approximate linear expression (3.4) for curvature and accurate nonlinear expression (3.3) for load equilibrium into Eq. (3.1) produces the following beam governing differential equation:

$$EI_{ZZ}U_Y''(X) = M_{ZL} + F_{YL}(1 + U_{XL} - X) - F_{XL}(U_{YL} - U_Y(X)) \tag{3.10}$$

This equation may be solved in closed-form by differentiating it twice with respect to X:

$$U_Y^{iv}(X) = \frac{F_{XL}}{EI_{ZZ}}U_Y''(X) \tag{3.11}$$

and applying the following four boundary conditions:

$$U_Y(0) = 0, \quad U_Y'(0) = 0, \quad U_Y''(L) = \frac{M_{ZL}}{EI_{ZZ}}, \quad U_Y'''(L) = \frac{-F_{YL} + F_{XL}U_Y'(L)}{EI_{ZZ}} \tag{3.12}$$

The importance of applying load equilibrium in the deformed configuration of the beam is that while the axial direction load F_{XL} finds a place in this differential equation, the equation itself and associated boundary conditions remain linear in the transverse-direction loads (F_{YL} and M_{ZL}) and displacements ($U_Y(X)$ and its derivatives). Consequently, solving this equation leads to linear relations between these end-loads and end-displacements (U_{YL} and $\theta_{ZL} = U'_{YL}$). The associated stiffness terms, however, are no longer merely elastic terms, but transcendental functions of the axial load F_{XL}. These functions are expanded as an infinite series in F_{XL} and truncated to its first power, with less than 1% error when F_{XL} is comparable to the transverse loads F_{YL} and M_{ZL}/L, to yield the following transverse end load–displacement relation:

$$\begin{bmatrix} F_{YL}L^2/EI_{ZZ} \\ M_{ZL}L/EI_{ZZ} \end{bmatrix} = \begin{bmatrix} k_{11}^{(0)} & k_{12}^{(0)} \\ k_{12}^{(0)} & k_{22}^{(0)} \end{bmatrix} \begin{bmatrix} \dfrac{U_{YL}}{L} \\ \theta_{ZL} \end{bmatrix} + \dfrac{F_{XL}L^2}{EI_{ZZ}} \begin{bmatrix} k_{11}^{(1)} & k_{12}^{(1)} \\ k_{12}^{(1)} & k_{22}^{(1)} \end{bmatrix} \begin{bmatrix} \dfrac{U_{YL}}{L} \\ \theta_{ZL} \end{bmatrix} \qquad (3.13)$$

Next, the geometric constraint imposed by the beam arc length may be captured via the following integral, to determine the dependence of the axial displacement U_{XL} on the transverse displacements:

$$L + \frac{1}{k_{33}} \frac{F_{XL}L^3}{EI_{ZZ}} = \int\limits_{0}^{L+U_{XL}} \left\{ 1 + \frac{1}{2} \left(U'_Y(X) \right)^2 \right\} dX \qquad (3.14)$$

The left- and right-hand sides of this equation represent the beam length before and after the bending deflection, respectively. The undeformed beam length is augmented with any elastic stretch resulting from the applied axial load F_{XL} on the left-hand side. In this case, it is important to include the second-order term in $U'_Y(X)$ on the right-hand side to capture the kinematics associated with the beam deflection geometry, and consistent with applying load equilibrium in the deformed condition (3.3).

Using the $U_Y(X)$ solution for Eq. (3.11), Eq. (3.14) may also be solved in closed form to reveal a component of U_{XL} that has a quadratic dependence on U_{YL} and θ_{ZL}. As might be expected, the coefficients in this quadratic relation are also transcendental functions of the axial load F_{XL}. A series expansion and truncation to the first power in F_{XL} yields:

$$\begin{aligned} \frac{U_{XL}}{L} = \frac{1}{k_{33}} \frac{F_{XL}L^2}{EI_{ZZ}} &+ \begin{bmatrix} \dfrac{U_{YL}}{L} & \theta_{ZL} \end{bmatrix} \begin{bmatrix} g_{11}^{(0)} & g_{12}^{(0)} \\ g_{12}^{(0)} & g_{22}^{(0)} \end{bmatrix} \begin{bmatrix} \dfrac{U_{YL}}{L} \\ \theta_{ZL} \end{bmatrix} \\ &+ \frac{F_{XL}L^2}{EI_{ZZ}} \begin{bmatrix} \dfrac{U_{YL}}{L} & \theta_{ZL} \end{bmatrix} \begin{bmatrix} g_{11}^{(1)} & g_{12}^{(1)} \\ g_{12}^{(1)} & g_{22}^{(1)} \end{bmatrix} \begin{bmatrix} \dfrac{U_{YL}}{L} \\ \theta_{ZL} \end{bmatrix} \end{aligned} \qquad (3.15)$$

For convenience of discussion, the three terms on the RHS above may be individually identified as $U_{XL}^{(e)}$, $U_{XL}^{(k)}$, and $U_{XL}^{(e-k)}$, respectively, and will be further described shortly.

Table 3.1 Characteristic coefficients for a simple beam

$k_{11}^{(0)}$	12	$k_{11}^{(1)}$	6/5	$g_{11}^{(0)}$	$-3/5$	$g_{11}^{(1)}$	1/700	
$k_{12}^{(0)}$	-6	$k_{12}^{(1)}$	$-1/10$	$g_{12}^{(0)}$	$1/20$	$g_{12}^{(1)}$	$-1/1400$	$k_{33} = \dfrac{12}{(T/L)^2}$
$k_{22}^{(0)}$	4	$k_{22}^{(1)}$	2/15	$g_{22}^{(0)}$	$-1/15$	$g_{22}^{(1)}$	11/6300	

Equations (3.13) and (3.15) constitute the beam constraint model (BCM) and provide accurate, compact, closed-form, and parametric relations between the end-loads and end-displacements of a simple beam. Further, in this format, all loads, displacements, and stiffness terms are naturally normalized with respect to the beam parameters: displacements and lengths are normalized by the beam length L, forces by EI_{ZZ}/L^2, and moments by EI_{ZZ}/L. Thus, one may define:

$$\frac{F_{XL}L^2}{EI_{ZZ}} \triangleq f_{x1}; \quad \frac{F_{YL}L^2}{EI_{ZZ}} \triangleq f_{y1}; \quad \frac{M_{ZL}L}{EI_{ZZ}} \triangleq m_{z1}$$

$$\frac{U_{XL}}{L} \triangleq u_{x1}; \quad \frac{U_{YL}}{L} \triangleq u_{y1}; \quad \theta_{ZL} \triangleq \theta_{z1}; \quad \frac{T}{L} \triangleq t; \quad \frac{X}{L} \triangleq x \qquad (3.17)$$

In the rest of this chapter, lower case symbols are used to represent normalized variables and parameters, as per the above convention. It has been shown that the stiffness coefficients k terms and constraint coefficients g terms, in general, are nondimensional *beam characteristic coefficients* that are solely dependent on the beam shape and not its actual size. These coefficients take the numerical values shown in Table 3.1 for a simple beam with uniform thickness along its length.

The BCM helps characterize the constraint behavior of a simple beam flexure in terms of its stiffness and error motions. Error motions are the undesired motions in a flexure element or mechanism: any motion in a DoF direction, other than the intended DoF, is referred to as cross-axis coupling, and any motion along a DoC direction is referred to as parasitic error. The first matrix term on the right-hand side of Eq. (3.13) represents the linear elastic stiffness in the DoF directions, analogous to Eq. (3.7). The second matrix on the right-hand side of Eq. (3.13) captures load stiffening (also known as geometric stiffening), which highlights the change in the effective stiffness in the DoF directions due to a DoC force. Both these matrix terms also capture the cross-axis coupling between the two DoF.

Equation (3.15) shows that the DoC direction displacement, which is a parasitic error motion, comprises three terms. $u_{x1}^{(e)}$ is a purely elastic component resulting from the stretching of the beam neutral axis in the X-direction, which is analogous to result of Eq. (3.8). $u_{x1}^{(k)}$ represents a purely kinematic component dependent on the two DoF displacements, and arises from the constant beam arc-length constraint. $u_{x1}^{(e-k)}$ represents an elastokinematic component, so called because of its elastic dependence on the DoC force f_{x1} and its kinematic dependence on the two DoF displacements. The elastokinematic component is also a consequence of the beam arc-length constraint, and arises due to a change in the beam deformation when f_{x1} is applied, even as u_{y1} and θ_{z1} are held fixed. The kinematic component $u_{x1}^{(k)}$ dominates the error motion in this DoC direction and increases quadratically with increasing DoF displacements.

The elastokinematic component of the DoC displacement, while small with respect to the purely kinematic component, is comparable to the purely elastic component and causes the DoC direction compliance to increase quadratically (and stiffness to decrease) from its nominal linear elastic value with increasing DoF displacements.

Thus, the BCM not only highlights the nonideal constraint behavior of a beam flexure, it also reveals interdependence and fundamental tradeoffs between the DoF quality (large range, low stiffness) and DoC quality (high stiffness, low parasitic error). The beam characteristic coefficients serve as beam-shape optimization parameters in flexure mechanism design. Moreover, the BCM accommodates any generalized end-load and end-displacement conditions in a scale-independent, compact, and parametric format.

As seen above, the application of load equilibrium in a deformed configuration to include the contribution of the axial force proves to be crucial in the constraint characterization of a beam. In the case of a clamped-clamped beam, its consequence is significant even for DoF displacements, u_{y1} and θ_{z1}, less than 0.1. However, in spite of including this nonlinear effect, the beam-governing differential equation remains linear in transverse loads and displacements, leading to a relatively simple mathematical model. On the other hand, relaxing the beam-curvature linearization assumption neither offers additional insights in constraint behavior, nor does its effect become significant until the DoF displacements are greater than 0.1. Yet, it renders the beam-governing equation nonlinear and therefore unusable for closed-form analysis. The BCM assumptions are carefully chosen such that they capture only the relevant nonlinearities, thus providing accuracy within a practical load and displacement range, and yet do not make the model unwieldy.

We next proceed to provide a comparison between the BCM for a simple beam and the corresponding full nonlinear FEA in ANSYS. Figure 3.3 plots the elastic stiffness coefficients ($k_{11}^{(0)}$, $k_{12}^{(0)}$, and $k_{22}^{(0)}$) and load-stiffening coefficients ($k_{11}^{(1)}$, $k_{12}^{(1)}$, and $k_{22}^{(1)}$) versus

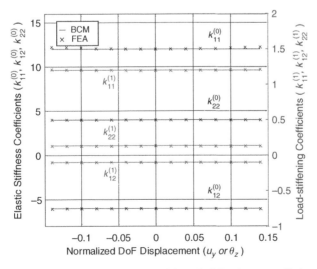

Figure 3.3 Elastic stiffness coefficients and load-stiffening coefficients for a simple beam: BCM versus FEA

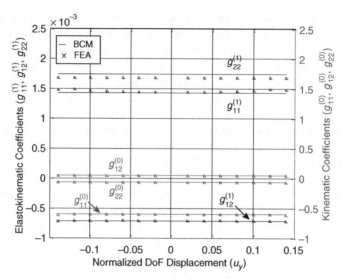

Figure 3.4 Kinematic and elastokinematic coefficients for a simple beam: BCM versus FEA

the normalized DoF displacement u_{y1} or θ_{z1}. Similarly, Figure 3.4 plots the kinematic $(g_{11}^{(0)}, g_{12}^{(0)},$ and $g_{22}^{(0)})$ and elastokinematic $(g_{11}^{(1)}, g_{12}^{(1)},$ and $g_{22}^{(1)})$ coefficients.

3.4 Case Study: Parallelogram Flexure Mechanism

Next, the effectiveness of the BCM in accurately predicting the motion guidance attributes of flexure mechanisms is highlighted using the parallelogram flexure mechanism (Figure 3.5), comprised of two identical simple beams ($L = 250$ mm, $T = 5$ mm, $H = 50$ mm, $W = 75$ mm, $E = 210\,000$ N mm^{-2}).

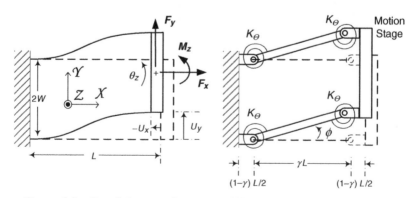

Figure 3.5 Parallelogram flexure and its pseudo-rigid body model

Using the normalization convention introduced earlier, the linear model for this flexure mechanism may be shown to be:

$$u_y = \frac{f_y}{24}; \quad u_x = \frac{t^2}{24}f_x; \quad \theta_z = \frac{t^2}{24w^2}\left[m_z + \frac{f_y}{2}\right] \tag{3.17}$$

The nonlinear load–displacement results for this flexure mechanism can be derived based on the BCM, using either explicit or energy methods (see reference [3] under the Further Reading section):

$$
\begin{aligned}
u_y &= \frac{f_y}{\left(2k_{11}^{(0)} + k_{11}^{(1)}f_x\right)} \\
u_x &= \frac{t^2}{24}f_x + g_{11}^{(0)}u_y^2 + \frac{g_{11}^{(1)}}{2}u_y^2 f_x \\
\theta_z &= \frac{1}{2w^2}\left(\frac{1}{k_{33}} + g_{11}^{(1)}u_y^2\right)\left[m_z - f_y\frac{\left(2k_{12}^{(0)} + k_{12}^{(1)}f_x\right)}{\left(2k_{11}^{(0)} + k_{11}^{(1)}f_x\right)}\right]
\end{aligned}
\tag{3.18}
$$

In the above relations, all loads and displacements are normalized as per Eq. (3.16). Beam thickness and beam-spacing dimensions are normalized as: $t = T/L$ and $w = W/L$. Substituting the values of the beam characteristic coefficient for a simple beam from Table 3.1, these relations reduce to:

$$
\begin{aligned}
u_y &= \frac{f_y}{(24 + 1.2f_x)} \\
u_x &= \frac{1}{2k_{33}}f_x - \frac{3}{5}u_y^2 + \frac{1}{1400}u_y^2 f_x \\
\theta_z &= \frac{1}{2w^2}\left(\frac{t^2}{12} + \frac{u_y^2}{700}\right)\left[m_z + u_y(12 + 0.1f_x)\right]
\end{aligned}
\tag{3.19}
$$

A PRBM is also illustrated alongside the parallelogram flexure module in Figure 3.5. Assuming m_z and f_x to be zero, the model parameters are given by $\gamma = 0.8517$ and $k_\Theta = 2.65$, and the load–displacement results are given by:

$$f_y\cos\phi - f_x\sin\phi = 8k_\Theta\phi; \quad u_y = \gamma\sin\phi; \quad u_x = \gamma(\cos\phi - 1) \tag{3.20}$$

Since the actual loading conditions of the individual beams change as the mechanism is displaced, the PRBM parameters should ideally be updated with each incremental displacement step. However, this change in model parameters is assumed to be negligible.

Clearly, the Y (transverse)-direction represents a DoF, while the X (axial) and Θ_Z (transverse)-directions represent DoC, in this case. Key constraint behavior predictions made by the above three models along with results from a nonlinear FEA are

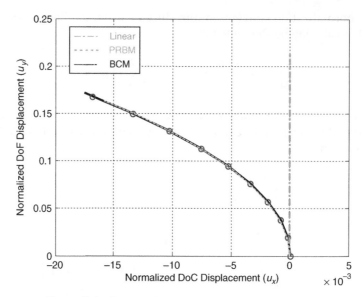

Figure 3.6   Dependence of u_x (DoC) on u_y (DoF)

plotted in Figures 3.6–3.8 over a u_y range of ± 0.15. Figure 3.6 plots the nonlinear dependence of u_x (X DoC parasitic error motion) on u_y (Y DoF displacement) and illustrates that both the PRBM and BCM capture the kinematic effect in beams very accurately. Figure 3.7 plots the variation in the X DoC stiffness with u_y (Y DoF displacement). While the PRBM does not recognize any compliance in this DoC direction

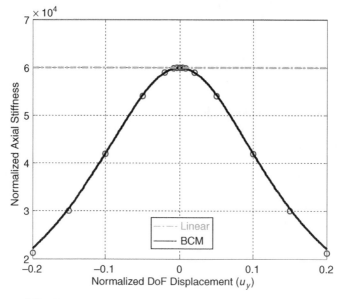

Figure 3.7   Dependence of X-direction (DoC) stiffness on u_y (DoF)

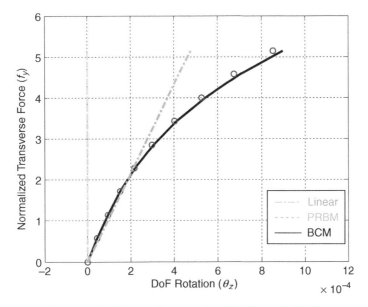

Figure 3.8   Dependence of θ_z (DoC) on $\boldsymbol{f_y}$ (DoF)

whatsoever, the linear model only captures the purely elastic stiffness component. On the other hand, the BCM accurately predicts the elastokinematic effects, as verified by the FEA. This variation in DoC stiffness has serious design consequences in terms of load bearing capacity and dynamic performance. Figure 3.8 plots θ_z (Θ_Z DoC parasitic error motion) with increasing f_y (Y DoF force). The PRBM predicts zero yaw rotation of the rigid stage, while the linear model is valid only for small forces and displacements. The BCM accurately captures this parasitic error motion, also influenced by the elastokinematic effect, even for large values of the DoF force and displacement.

3.5 Conclusions

The above case-study demonstrates the applicability and accuracy of the BCM. In particular, it is noteworthy that the model captures the stiffness and error motions associated with flexure mechanism being considered in a closed form and parametric manner for any general loading at the rigid motion stage. Moreover, the final load–displacement relations remain valid irrespective of the actual beam shapes. If the beams assume any shape other than the uniform thickness case considered here, the only difference will be in the beam characteristic coefficients. Also, any number of beams may be considered instead of the two considered here; the mathematical steps involved in the derivation of the final load–displacement relations remains the same.

The final results (3.18) highlight the performance trade-offs, qualitatively as well as quantitatively, that exist between the various desired attributes of the parallelogram flexure mechanism. For example,

1. While it is desirable to maximize the Y DoF displacement, the model shows that the X and Θ DoC error motions grow with the Y DoF displacement, highlighting the deviation for straight-line motion characteristics.
2. The X DoC error motion is dependent on the kinematic beam characteristic coefficient $g_{11}^{(0)}$ which can be shown to always remain nonzero irrespective of the beam shape. This makes physical sense because the X-direction displacement is the consequence of arc-length conservation, which is fundamental to the beam deformation kinematics.
3. The Θ DoC error motion exhibits a linear as well as cubic dependence on the Y DoF displacement. While the former arises from the linear elastic compliance of the beams in the axial direction, the latter is due to their nonlinear elastokinematic compliance. The elastokinematic compliance is dependent on the $g_{11}^{(1)}$ coefficient, which may be reduced via beam-shape optimization.
4. High stiffness along the X DoC is important because this is the load-bearing direction. However, the analytical results show that the compliance in this direction increases, from its nominal linear value, quite significantly with increasing Y DoF displacement. Once again, this dependence is based on the elastokinematic coefficient $g_{11}^{(1)}$, which may be reduced via beam-shape optimization. This can help reduce the rate at which the X DoC compliance increases and stiffness drops, with increasing Y DoF displacement.
5. In general, it is desirable to reduce the Y DoF stiffness in motion guidance applications. The above analytical results show load stiffening in the Y DoF direction in the presence of a force in the X DoC. The BCM elucidates that this behavior arises from the beam characteristic coefficient $k_{11}^{(1)}$, which is also fundamental to the beam deformation kinematics and may not be reduced much via beam shape optimization. This shows that the load-stiffening effect cannot be ignored in the parallelogram flexure, when displacements are intermediate or large. In fact, this effect may be exploited to reduce the Y DoF stiffness by applying a compressive X DoC force.

Although one representative application of the BCM is shown here, several other variations and applications are covered in the Further Reading suggestions below.

Further Reading

[1] Jones, R.V., 1988, *Instruments and Experiences: Papers on Measurement and Instrument Design*, John Wiley & Sons, New York, NY.
[2] Smith, S.T., 2000, *Flexures: Elements of Elastic Mechanisms*, Gordon and Breach Science Publishers, New York, NY.
[3] Awtar, S., Slocum, A.H., and Sevincer, E., 2006, "Characteristics of Beam-based Flexure Modules", *ASME Journal of Mechanical Design*, **129**(6), pp 625–639.

[4] Awtar, S., and Sen, S., 2010, "A Generalized Constraint Model for Two-dimensional Beam Flexures: Non-linear Load-Displacement Formulation", *ASME Journal of Mechanical Design*, **132**(8), pp. 0810081-08100811.
[5] Awtar, S., and Sen, S., 2010, "A Generalized Constraint Model for Two-dimensional Beam Flexures: Non-linear Strain Energy Formulation", *ASME Journal of Mechanical Design*, **132**(8), pp. 0810091-08100911.
[6] Sen, S. and Awtar, S., 2012, "A Closed-Form Nonlinear Model for the Constraint Characteristics of Symmetric Spatial Beams," ASME Journal of Mechanical Design, in press.

4

Modeling of Large Deflection Members

Brian Jensen

Brigham Young University, USA

4.1 Introduction

The previous chapter developed methods to model compliant mechanisms with small or intermediate-range deflections (up to about 10% of the length of a beam). For many compliant mechanisms, such an analysis provides useful insights into the behavior of the system, and gives readily used modeling tools. However, the analysis may be inappropriate or inexact when flexible beams experience large deflections. This may occur when a mechanism's desired motion is comparable to its size, or when a nonlinear force–deflection relationship is needed, as in a bistable mechanism. In these cases, the assumptions made during small- or intermediate-deflection analysis can lead to significant errors, and models designed for large deflections should be used.

For example, this chapter shows how to use large-deflection modeling on fixed-pinned beams (element EM-1 in the handbook library), as well as fixed-guided beams (element EM-4). These elements are then used in a wide variety of mechanisms, including several bistable mechanism designs (mechanisms M-9 through M-12), as well as straight-line suspension mechanisms (such as mechanisms M-15 through M-18, M-28, M-38, M-39, and M-79).

The classical tool used to solve for large deflections is elliptic integrals, a class of functions that arise in the solution of differential equations for large beam deflections [1]. More recently, nonlinear finite element modeling [2–4], and direct numerical integration of the differential equations for large deflections [5], have provided accurate prediction of large-deflection behavior. Nonlinear finite element modeling, in particular, provides accurate solutions for an extremely wide range of problems. However,

Handbook of Compliant Mechanisms, First Edition. Edited by Larry L. Howell, Spencer P. Magleby and Brian M. Olsen.
© 2013 John Wiley & Sons, Ltd. Published 2013 by John Wiley & Sons, Ltd.

in problems requiring solutions for buckled beams, which frequently arise in the design of bistable mechanisms, the finite element method does not yet predict the buckling mode accurately, leading to large errors in the solutions [2]. Moreover, in the early design stage, rapid modeling of a large number of potential solutions is often desirable. Elliptic integral solutions can provide this rapid feedback to aid in selecting an appropriate design, which may then be further optimized using finite element models. This chapter demonstrates how such cases may be modeled using elliptic integrals.

4.2 Equations of Bending for Large Deflections

Figure 4.1 shows an initially straight beam of length L with constant cross section. The beam's material has Young's modulus E, and the cross section has second moment of area I. Large-deflection analysis results in a coupled set of three non-dimensional equations:

$$\sqrt{\alpha} = F\left(k, \phi_2\right) - F\left(k, \phi_1\right) \tag{4.1}$$

$$\frac{b}{L} = -\frac{1}{\sqrt{\alpha}}\left\{2k\cos\psi\left(\cos\phi_1 - \cos\phi_2\right) + \sin\psi\left[2E\left(k, \phi_2\right) - 2E\left(k, \phi_1\right)\right.\right.$$
$$\left.\left. - F\left(k, \phi_2\right) + F\left(k, \phi_1\right)\right]\right\}. \tag{4.2}$$

$$\frac{a}{L} = -\frac{1}{\sqrt{\alpha}}\left\{2k\sin\psi\left(\cos\phi_2 - \cos\phi_1\right) + \cos\psi\left[2E\left(k, \phi_2\right) - 2E\left(k, \phi_1\right)\right.\right.$$
$$\left.\left. - F\left(k, \phi_2\right) + F\left(k, \phi_1\right)\right]\right\}. \tag{4.3}$$

Here, many of the variables are defined in Figure 4.1. In addition, α is the nondimensional force given by

$$\alpha = \frac{RL^2}{EI} \tag{4.4}$$

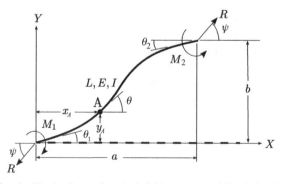

Figure 4.1 An illustration of a straight beam and its deflected shape

The functions $F(k, \phi)$ and $E(k, \phi)$ are the elliptic integrals of the first and second kind, respectively [6]. These functions may be thought of as similar to the trigonometric functions sine and cosine. As with the trigonometric functions, the elliptic integral functions can be very rapidly evaluated using numerical methods on a computer. The nondimensional parameter k, known as the modulus of the function, can vary between 0 and 1. In this application, k corresponds roughly but nonlinearly to the magnitude of the force R. The variable ϕ, with units of radians, is called the amplitude of the elliptic integral. It varies continuously along the beam from ϕ_1 on the left edge to ϕ_2 on the right. It is related to the beam angle θ by the relation

$$k \sin \phi = \cos \left(\frac{\psi - \theta}{2} \right) \tag{4.5}$$

where, for ϕ_1 or ϕ_2, the angles θ_1 or θ_2 are used. In addition, the beam's end moments are given by

$$M_{1,2} = 2k\sqrt{EIR} \cos \phi_{1,2} \tag{4.6}$$

Equations (4.1)–(4.3) are the key equations used in nonlinear beam analysis. Essentially, Eq. (4.1) describes the force acting on the end of the beam, and Eqs. (4.2) and (4.3) describe the horizontal and vertical deflections of the end of the beam. For the full derivation of these equations, see [7] or [8].

Solutions to these equations normally require nonlinear numerical solution, depending on the boundary conditions for the beam. Once the equations are solved, the full deflected shape of the beam can also be found. For an arbitrary point A along the beam, its deflected x- and y-coordinates are given by

$$\frac{x_A}{L} = -\frac{1}{\sqrt{\alpha}} \left\{ \cos \psi \left[2E(k, \phi) - 2E(k, \phi_1) - F(k, \phi) + F(k, \phi_1) \right] \right.$$
$$\left. + 2k \sin \psi (\cos \phi - \cos \phi_1) \right\}. \tag{4.7}$$

$$\frac{y_A}{L} = -\frac{1}{\sqrt{\alpha}} \left\{ \sin \psi \left[2E(k, \phi) - 2E(k, \phi_1) - F(k, \phi) + F(k, \phi_1) \right] \right.$$
$$\left. + 2k \cos \psi (\cos \phi_1 - \cos \phi) \right\}. \tag{4.8}$$

for an arbitrary value ϕ that is between ϕ_1 and ϕ_2. The distance s along the beam to point A is

$$s = \sqrt{\frac{EI}{R}} \left[F(k, \phi) - F(k, \phi_1) \right] \tag{4.9}$$

4.3 Solving the Nonlinear Equations of Bending

Most commonly, two basic approaches are used for solving Eqs. (4.1)–(4.3). In what might be called the "forward" solution approach, the applied forces are known, and the beam deflection must be found. In the "reverse" approach, the beam deflections are known, and the applied forces must be found. In either case, the equations are readily solved using a nonlinear solver.

The next section gives two examples of solutions in common compliant mechanism problems: the bending of a fixed-pinned beam and the motion of a bistable mechanism consisting of a tilted fixed-guided beam. The first example shows the forward solution approach, while the second shows the reverse solution approach.

4.4 Examples

4.4.1 Fixed-Pinned Beam

Fixed-pinned beams often arise in partially compliant mechanisms. Figure 4.2 shows an illustration of a beam fixed on the left edge and pinned on the right. The beam is loaded by an end force of magnitude R and direction ψ, and the problem is to find a and b, the coordinates of the end of the deflected beam.

Based on the stated boundary conditions, the left end angle θ_1 (shown in Figure 4.1) is 0, and the right end moment M_2 (also shown in Figure 4.1) is 0. The problem then becomes one of finding the value of k that satisfies Eq. (4.1), given that ϕ_1 must satisfy Eq. (4.5) at the fixed end of the beam, and ϕ_2 must satisfy Eq. (4.6) at the free end of the beam. This second condition gives $\phi_2 = \pi/2$, since $M_2 = 0$. Any desired method can be used to solve Eq. (4.1), including the Newton–Raphson method or a bounded method such as the false position method. Then, once k is known, Eqs. (4.2) and (4.3) can be solved directly to give the horizontal and vertical deflections of the end of the beam, or Eqs. (4.7) and (4.8) can be solved to give the full deflected shape of the beam. Similarly, Eq. (4.5) can be solved to find θ_2, the angle of the deflected end of the beam.

Code is posted online at http://compliantmechanisms.byu.edu/content/downloads giving two ways to model fixed-pinned segments. The Excel spreadsheet fixed-pinnedbending.xls uses visual basic macros for the elliptic integrals to solve this problem. Similarly, the MATLAB script fpbending.m shows how to solve this problem using the function fpbeambending.m. The script solves for the deflection

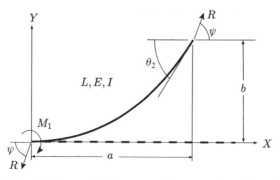

Figure 4.2 A fixed-pinned beam (with no end moment on the right) shown in the initially straight and deflected states

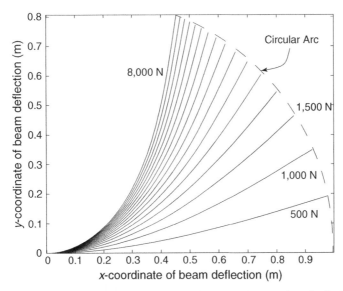

Figure 4.3 Deflected beam shapes for a fixed-pinned beam loaded with a vertical force with magnitude varying from 500 to 8000 N. For comparison, a circular arc is shown to match well the end deflection. The *x*- and *y*-axes are drawn at the same scale

of a steel beam (with Young's modulus of 200 GPa) 1 m long, with a thickness of 1 cm, and a width of 5 cm. The beam is loaded with a vertical force ($\psi = 90°$) ranging from 500 N to 8000 N, in steps of 500 N. The resulting beam deflections are shown in Figure 4.3. Two features are especially noted. First, the deflected path of the beam's pinned end is approximately circular, with a center midway along the length of the straight beam (though closer to the fixed end than the middle of the beam). Secondly, the beam effectively becomes stiffer as the force increases. For small forces (near 500–1500 N), additional force causes large additional deflection. For larger forces (near the top of the range shown), additional force causes much smaller additional deflections.

The first feature represents the genesis of the pseudo-rigid-body model concept, described more fully in Chapter 5. In the pseudo-rigid-body model, the motion of a compliant beam is represented by a rigid link pinned to a second rigid link. This concept provides significant reduction in the effort required to model many compliant mechanisms. The second feature is also captured in the pseudo-rigid-body model due to the reduction in effective moment acting on the model as horizontal deflection increases.

4.4.2 Fixed-Guided Beam (Bistable Mechanism)

A bistable mechanism may be created by opposing two banks of angled beams against a central shuttle, as shown in Figure 4.4. In this mechanism, each beam may be modeled as a fixed-guided beam, with the central shuttle free to move vertically

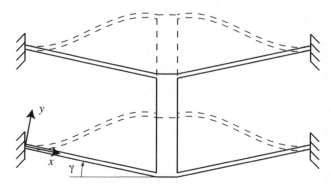

Figure 4.4 A bistable mechanism shown in two positions that uses four fixed-guided beams. Each beam may be modeled using Eqs. (4.1)–(4.3)

while constraining each beam to have no end deflection ($\theta_2 = 0$). By rotating the coordinate system so that the x-axis lies along the initially straight beam, the beam can by modeled as experiencing deflections along a line rotated with respect to the vertical, as shown in Figure 4.5. In this case, the deflections are known to occur along this line (the load line), and the forces required to create those deflections are to be found. Hence, this system uses the reverse solution approach.

The solution is somewhat more difficult because the unknowns, which are k, R, and ψ, are all found in Eqs. (4.2) and (4.3). Hence, Eqs. (4.1), (4.2), and (4.3) must all be solved simultaneously for these three unknowns. For a given guess of ψ and k, Eq. (4.5) can be solved for ϕ_1 and ϕ_2. However, because $\theta_1 = \theta_2 = 0$, both cases result in the equation

$$\sin \phi_{1,2} = \frac{1}{k} \cos \left(\frac{\psi}{2} \right) \tag{4.10}$$

Hence, unique solutions require that ϕ_1 be the principal solution to Eq. (4.10), while ϕ_2 be a higher-order solution. This gives rise to different modes of solutions, with the first mode given by

$$\phi_2 = \pi - \phi_1 \tag{4.11}$$

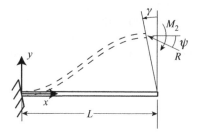

Figure 4.5 Each beam in the bistable mechanism may be modeled as a fixed-guided beam under deflection along a line rotated an angle γ with respect to the vertical

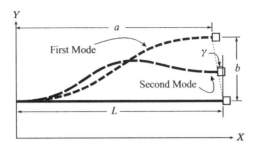

Figure 4.6 First and second mode shapes for a fixed-guided beam

and the second mode given by

$$\phi_2 = \phi_1 + 2\pi \tag{4.12}$$

Higher-order mode solutions are also possible, but these will not be seen in practice because they are statically unstable. The first-mode solution represents deflections with a single inflection point, while the second-mode solution represents deflections with two inflection points, as shown in Figure 4.6. For a given horizontal and vertical deflection of the end of the beam, a solution can be found using one and only one of these two modes, with each one representing different regions in the beam's deflected space. For more information on this topic, see [8].

The MATLAB script fixedguidedbeambending.m, found online at http://compliantmechanisms.byu.edu/content/downloads, solves this problem. Figure 4.7

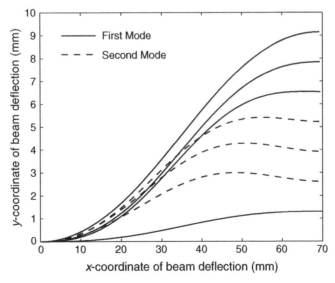

Figure 4.7 Several deflected beam shapes from the solution for the bistable beam. The axes are not drawn at the same scale

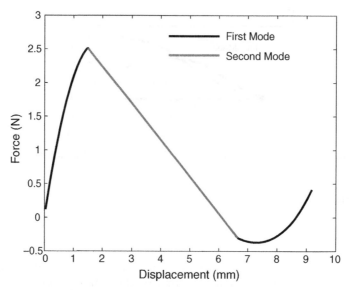

Figure 4.8 Force–deflection curve for a bistable beam

shows several beam solutions along the load line for a bistable beam with length 70 mm, inplane thickness of 1.5 mm, and out-of-plane width of 12 mm. The beam's load line is tilted at an angle of 5°, and it is constructed from polypropylene with a Young's modulus of 1.4 GPa. Notice that the smallest deflections are in the first mode, followed by a region in the second mode, followed again by first-mode solutions. Figure 4.8 shows the corresponding force–deflection data for a force directed along the load line. (Note that the force perpendicular to the load line will be counteracted by an equal and opposite force from the other side of the bistable mechanism.) The curve shows bistable behavior, with an unstable equilibrium position at 6.1 mm and a second stable position at 8.6 mm. The figure shows the initial region of first-mode bending, the middle region of second-mode bending, and the final transition to first-mode bending before the second stable position is reached. These features are common to all bistable beams with the general form shown in Figure 4.4. Another interesting feature of the force–displacement curve is that the force in the second-mode region is a nearly straight line with negative stiffness. For this reason, these mechanisms have also been proposed for force compensation in statically balanced mechanisms [9, 10].

Note that nonlinear finite element solvers tend to incorrectly predict the buckling mode for this case, resulting in erroneous solutions [2, 8]. While carefully-applied tricks can be used to correct the solutions, the method shown here produces accurate solutions without problems.

4.5 Conclusions

The methods shown in this chapter, along with the accompanying code, show how to use elliptic integral solutions to model compliant beams with large deflections. While these solutions are not as easy and straightforward as the small- and medium-range deflection models shown in the previous chapter, they provide a strong design tool for beams with large, highly nonlinear deflections. In particular, elliptic integral solutions are especially strong in predicting motion of buckled beams used in compliant bistable mechanisms, for which straightforward finite element models are not accurate. The method shown in this chapter readily predicts the first- and second-mode bending deflections that arise for fixed-guided beams used in compliant bistable mechanisms. The chapter also showed an example of analyzing the deflection of a fixed-pinned beam, which gives rise to the concept of a pseudo-rigid-body model (discussed further in Chapter 5).

Further Reading

For more information on large-deflection modeling, a number of sources are recommended. The book *Flexible Bars* by Frisch-Fay [1] develops in great detail the equations for modeling of beam deflections using elliptic integrals. The book emphasizes the mathematics of the solutions but touches on applications as well. The classic paper by Shoup and McLarnan [7] shows how these beam deflection equations can be applied to a practical compliant mechanism design. Larry Howell's text [11] also demonstrates numerous large-deflection solutions, and shows how they led to the formulation of the pseudo-rigid-body model. Some recent publications have also demonstrated the use of elliptic integral solutions in both modeling and design of compliant mechanisms [8, 12]. For more information specifically on the mathematics of elliptic integral functions, the classic handbook by Abramowitz and Stegun contains excellent reference material [6].

The modeling tools prepared to accompany this chapter use both Microsoft Excel and Mathworks MATLAB. Numerous references and tutorials exist to find more information about both. In particular, the Mathworks website at http://www.mathworks.com contains a large amount of information on using MATLAB, including several tutorials designed to fit several different learning styles.

References

[1] Frisch-Fay, R., *Flexible Bars*, Butterworth, Washington, D.C., 1962.
[2] Wittwer, J. W., Baker, M. S., and Howell, L. L., "Simulation, measurement, and asymmetric buckling of thermal microactuators," *Sensors and Actuators A: Physical*, Vol. 128, pp. 395–401, 2006.
[3] Shamshirasaz, M. and Asgari, M. B., "Polysilicon micro beams buckling with temperature-dependent properties," *Microsystem Technologies*, Vol. 14, pp. 975–961, 2008.
[4] Masters, N. D. and Howell, L. L. "A self-retracting fully compliant bistable micromechanism," *Journal of Microelectromechanical Systems*, Vol. 12, pp. 273–280, 2003.
[5] Zhao, J., Jia, J., He, X., and Wang, H., "Post-buckling and snap-through behavior of inclined slender beams," *ASME Journal of Applied Mechanics*, Vol. 75, p. 041020, 2008.
[6] Abramowitz, M. and Stegun, I. A., *Handbook of Mathematical Functions with Formulas, Graphs, and Mathematical Tables*, U.S. Government Printing Office, 1972.

[7] Shoup, T. E. and McLarnan, C. W. "On the use of the undulating elastica for the analysis of flexible link mechanisms," *Journal of Engineering in Industry*, Vol. 93, pp. 263–267, 1971.

[8] Holst, G. L, Teichert, G. H., and Jensen, B. D., "Modeling and experiments of buckling modes and deflection of fixed-guided beams in compliant mechanisms," *ASME Journal of Mechanical Design*, Vol. 133, p. 051002, 2011.

[9] Tolou, N., Henneken, V. A., and Herder, J. L., "Statically Balanced Compliant Micro Mechanisms (SB-MEMS): Concepts and Simulation," in *Proceedings of ASME 2010 International Design Engineering Technical Conference & Computers and Information in Engineering Conference*, paper no. DETC2010-28406, 2010.

[10] Tolou, N., Estevez, P., and Herder, J. L., "Collinear-Type Statically Balanced Compliant Micro Mechanism (SB-CMM): Experimental Comparison Between Pre-Curved and Straight Beams," in *Proceedings of ASME 2011 International Design Engineering Technical Conference & Computers and Information in Engineering Conference*, paper no. DETC2011-47678, 2011.

[11] Howell, L. L., *Compliant Mechanisms*, John Wiley & Sons, New York, 2001.

[12] Todd, B., Jensen, B. D., Shultz, S. M., and Hawkins, A. R., "Design and testing of a thin-flexure mechanism suitable for stamping from metal sheets," *ASME Journal of Mechanical Design*, Vol. 132, p. 071011, 2010.

5

Using Pseudo-Rigid Body Models

Craig Lusk

The University of South Florida, USA

5.1 Introduction

The purpose of this chapter is to: 1) describe why the pseudo-rigid-body model (PRBM) approach is useful, 2) provide a few 'rules of thumb' for using the PRBM approach, and 3) present illustrative examples. Pseudo-rigid-body models are useful for understanding the behavior of flexible parts and compliant mechanisms because they allow flexible bodies to be modeled as rigid bodies, thus allowing application of analysis and synthesis methods from rigid-body mechanisms such as those found in references [1–3].

PRBMs are a set of diagrams and equations that describe a correspondence between the motion and force of an elastic member and a rigid-body mechanism. The correspondence does not have to be exact, in order to be a useful analysis and design tool. Traditional modeling approaches for elastic bodies focus on stress and strain fields, i.e. point-by-point variations in force and displacement. PRBMs, on the other hand, describe the behavior of whole compliant segments, and hence are useful when tackling design issues at the component/device level.

Consider the partially compliant mechanism shown in Figure 5.1. It consists of three links that are pinned together. Links 1 and 2 are rigid. Link 3 consists of a rigid segment and a thin compliant segment. Since the links are pinned together, and assuming there were no flexibility in the third link, the arrangement would be an immobile structure. Yet, because there is flexibility, when an input torque is applied to link 2, the links can move. Furthermore, if the compliant segment is quite thin, it may undergo large enough deflections to invalidate the usual small deflection assumptions

Handbook of Compliant Mechanisms, First Edition. Edited by Larry L. Howell, Spencer P. Magleby and Brian M. Olsen.
© 2013 John Wiley & Sons, Ltd. Published 2013 by John Wiley & Sons, Ltd.

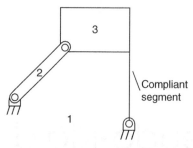

Figure 5.1 A partially compliant mechanism, composed of two rigid links (1 and 2) and a compound link (3) with a rigid rectangular segment and a compliant segment

of elementary beam theory. Thus, the device becomes quite complicated to model using a stress/strain approach. There is a mechanism equilibrium relationship that determines how the torque on link 2 transmits as a force to link 3. Simultaneously, there is a large deflection beam relationship that determines how the forces acting at the two pins on link 3 cause the compliant segment to deform. The relationship equations must be solved simultaneously, because the equilibrium requirement link 2 requires knowledge about the positions of the pins of link 3, and the beam deflection problem requires knowledge of the applied forces from link 2.

The PRBM solves this conundrum by modeling the compound link with two rigid links as shown in Figure 5.2. One of these links represents the rigid segment; the other, the pseudo-rigid-body link, represents the motion of the compliant segment. The rotation of link 2, and the rigid portion of link 3, can be simulated with a high degree of accuracy by choosing the appropriate location for the pin connecting the rigid portion of link 3, and the pseudo-rigid-body link. Additionally, it is possible to include the flexible segment's resistance to bending by including a torsional spring at the pseudo joint (i.e. the simulated joint) in link 3 as represented in Figure 5.2.

The crux of using a PRBM to model the behavior of a compliant mechanism lies in making *defensible* choices for the pseudo-rigid links that will represent the compliant links. A PRBM is defensible if its advantages in simplicity are greater than its disadvantages due to loss of accuracy. There is a pseudo-joint location such that the model link's end-deflection exactly matches the compliant beam's end-deflection for a given

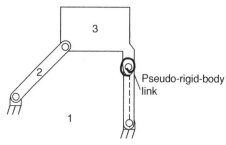

Figure 5.2 The pseudo-rigid-body model (PRBM) of the partially compliant mechanisms shown in Figure 5.1

loading condition. Yet, often, as in our example mechanism, the loading conditions change as link 2 moves. Thus, for simplicity, a single beam length is chosen that gives an accurate, but not exact, result for the motion range of the mechanism. In this particular example, the pseudo-rigid link is chosen to be 0.85 L, where L is the length of the compliant segment. The length of the pseudo-rigid link is called the *characteristic radius*. The value of 0.85 is known as the *characteristic radius factor* and is represented in equations with a Greek letter, γ. The torsional spring has a characteristic stiffness K, which is equal to 2.25 EI/L, where E is Young's modulus and I is the second moment of area of the segment [4].

5.2 Pseudo-Rigid-Body Models for Planar Beams

Planar beams, i.e. beams whose neutral axis lies in a plane, can exhibit a range of deflected shapes ranging from the perfectly straight to the perfectly circular. When a straight beam is loaded, it tends to become more curved, and when a circular beam is loaded it tends to straighten out. These two different extremes (and how they blend together) are captured precisely in the mathematical theory of beam bending. One of the key results that has been obtained from the combination of beam theory and PRBMs is that if a beam, after loading, has an inflection point the characteristic radius factor is between 0.83 and 0.85 [5].

A typical case of a planar beam is a fixed-free beam with a force on its end, as shown in Figure 5.3. In this model, the curvature of the beam is a minimum at the free end, and is a maximum at the fixed end. The motion of the beam under load is approximated by a PRBM, shown in Figure 5.4, by replacing the flexible beam with a rigid link that is pinned to a fixed link. The length of the fixed link is between 0.15L and 0.17L, and the length of the pseudo-rigid link is chosen to be between 0.83L and 0.85L. Under load, the pseudo-rigid beam rotates through an angle, Θ. For small deflections ($\Theta < 15°$), 0.83L is more accurate, for larger deflections ($\Theta > 45°$), 0.85L is

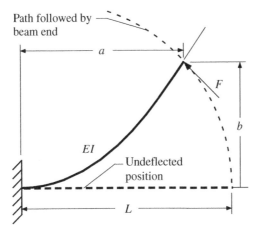

Figure 5.3 Fixed-free compliant beam with a force on the free end (4)

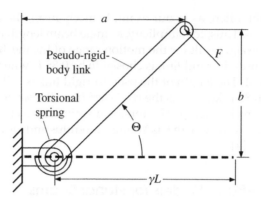

Figure 5.4 The PRBM of the fixed-free beam shown in Figure 5.3 (4)

better, but either will serve for initial design purposes [2]. The equations that describe the x- and y-coordinates (a and b, respectively) of the tip of the compliant beam are:

$$a = (1 - \gamma)L + \gamma L \cos \Theta \qquad (5.1)$$
$$b = \gamma L \sin \Theta \qquad (5.2)$$

The stiffness of this beam is captured by placing a torsional spring at the pseudo pivot. The collective resistance of the beam to being bent all along its length is then modeled in the PRBM, as the resistance located at the pseudo-pivot in a torsional spring with stiffness, K. The value of K is given by the following expression:

$$K = 2.25 \, EI/L \qquad (5.3)$$

where E is Young's modulus for the beam material, I is the second moment of area of the cross section, and L is the length of the compliant segment. Section A.1.3 provides additional information on the pseudo-rigid-body model for fixed-free beams.

Closely related to the fixed-free loading conditions are the fixed-guided beam (two curvature maximums and a minimum in the center), and a pinned-pinned buckled beam (two curvature minimums with a maximum in the center). These two are obtained from attaching two copies of the fixed-free beam shown in Figure 5.3 together, as shown in Figures 5.5 and 5.6. The two fixed-free beam models are each half of the length of the original fixed guided beam or pinned-pinned buckled beam. Likewise, the PRBMs for these configurations are obtained by attaching the two PRBMs of the flexible beam together as shown in Figures 5.7 and 5.8. In the fixed-guided beam, one end is fixed and the other is allowed to translate in the x- and y-directions, but is not allowed to rotate (thus, guided). This results in a symmetric loading pattern, with the maximum beam curvature (and stresses) at the ends of the beam. (See Section A.1.4 for more information on PRBMs for fixed-guided beams.) In the buckled pinned-pinned beam, the ends of the beam are allowed to rotate and thus no bending stress, or curvature occurs at the end of the beam, but compressive buckling loads are placed on the beam, resulting in a curvature (and stress) maximum in the center of the beam. (See Section A.1.7 for more on the pinned-pinned PRBM.)

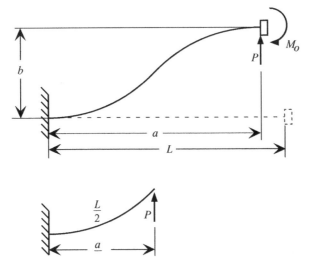

Figure 5.5   A fixed-guided beam. There is an inflection point (minimum curvature) in the center of the beam, and the beam is symmetric about that point (4)

A number of other interesting variations occur when loading conditions result in more than one point of maximum and/or minimum curvature. The key insight is that the pseudo-pivot does not occur at extreme values of curvature, but that if the distance between a given max/min pair on the beam is L, a pseudo-pivot occurs at a distance between $0.15L$ and $0.17L$ from the maximum and between $0.83L$ and $0.85L$ from the minimum. Appendix A of this chapter provides information on a number of pseudo-rigid-body models.

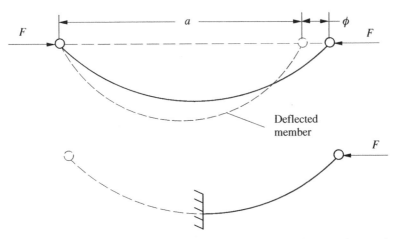

Figure 5.6   A buckled pinned-pinned beam. There is a maximum of curvature point in the center of the beam, and the beam is symmetric about that point. Adapted from (4)

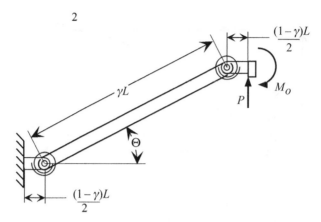

Figure 5.7 The PRBM of the fixed-guided beam, which joins together two copies of the PRBM of a fixed-free beam. Here, the middle segment is 0.85L and each of the side segments is 0.075L (4)

5.3 Using Pseudo-Rigid-Body Models: A Switch Mechanism Case-Study

The objective of this case study is to illustrate compliant mechanism design using the pseudo-rigid-body-models approach applied to different compliant mechanism designs. The mechanism chosen for this case study is a switch that has three distinct positions: forward, neutral, and reverse. The design calls for the mechanism to be in the neutral position when no load is applied; a force applied in the positive direction moves the switch to the forward position, and a force applied in the negative direction moves the switch to the reverse position.

The switch positions are shown in Figures 5.9a–c:

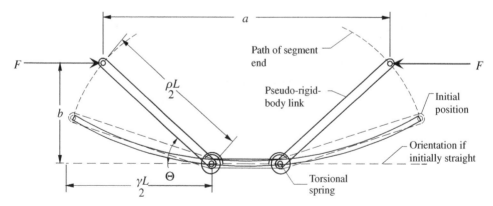

Figure 5.8 The PRBM of the buckled pinned-pinned beam, which joins together two copies of the fixed-free PRBM of Figure 5.6 with the point of maximum curvature in the middle. In this model, the middle segment is 0.15L and the side segments are each 0.425L. Adapted from (4)

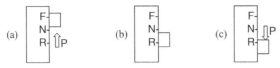

Figure 5.9 The three position of the compliant switch: Forward (F), Neutral (N), and Reverse (R). A force, P, is required to move the switch out of its neutral position

A variety of simple designs can provide the necessary motion and forces. A few will be considered for the purposes of this case study, including designs utilizing: 1) a small length flexural pivot, 2) a flexible beam, and 3) a fixed-guided beam, which are shown in Figures 5.10–5.15, respectively. The pseudo-rigid-body models for these different loading conditions are found in Appendix A.

Case I: The small-length flexural pivot (SLFP) switch

The SLFP switch consists of a slider (shown in light gray in Figure 5.10) and the switch body (shown in dark grey) which is connected to a spring element. In this case, the spring element consists of a small-length flexural pivot and a rigid segment. The kinematic pairs in the SLFP switch are a prismatic (or sliding) joint between the slider and the switch body, a sliding and rolling contact between the slider and the rigid segment (a higher pair), and the SLFP itself which, using the pseudo-rigid-body model, is treated as a rolling pair (i.e. a pin joint). The location of the characteristic pivot is taken to be the center of the small flexible link, and the stiffness of the flexible link is modeled with a torsional spring, $K = EI/L$. (See Section A.1.1 for more on SLFP PRBMs.) The motion of the switch can be found using the PRBM and standard mechanism analysis (vector loop) equations. The force required to deflect the beam can be found using Newton's laws for the two movable links in the PRBM.

Case II: The flexible beam switch

The flexible beam switch also consists of a slider, shown in light gray in Figure 5.12, and the switch body, shown in dark gray, which is connected to

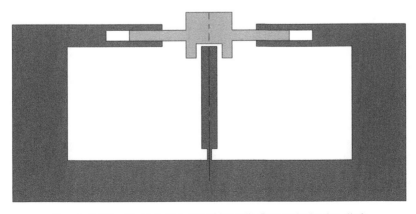

Figure 5.10 Case I: The small-length flexural pivot switch

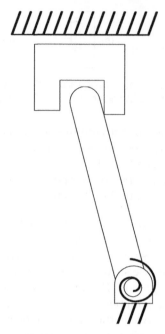

Figure 5.11   Case I PRBM

a spring element, which consists of a long flexible segment. The kinematic pairs in the flexible beam switch are identical to those found in the SLFP switch. The location of the characteristic pivot is $0.85L$, where L is the distance from the base of the flexure to the point of contact of the flexure and the slider. The stiffness of the torsional spring, $K = 2.25\ EI/L$. The motion of the switch can be found using the PRBM and standard mechanism analysis (vector loop) equations. The force required to deflect the beam can be found

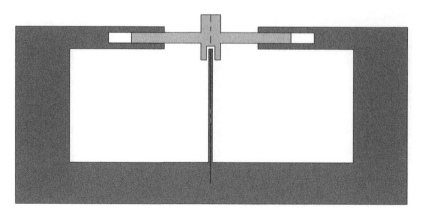

Figure 5.12   Case II: The flexible beam

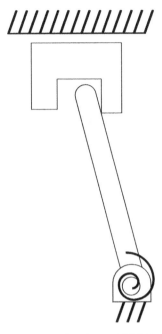

Figure 5.13   Case II PRBM

using Newton's laws for the pseudo-link and the slider. While the kinematic models of Case I and II switches are topologically identical, the stiffness and pivot location changes in Case II, because the location of the point of contact between the slider and flexure slides along the flexure as they move. This changing contact point changes the effective length of the PRBM link. Because the point moves toward the tip of the flexure as the slider deflects from its neutral position, it acts as a softening (decreasing stiffness) spring.

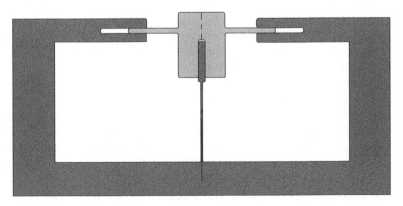

Figure 5.14   Case III: The fixed-guided beam

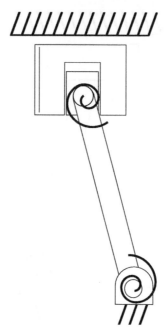

Figure 5.15   Case III PRBM

Case III: The fixed-guided beam switch

The flexible beam switch also consists of a slider, shown in light gray in Figure 5.14, and the switch body, shown in dark gray, which is connected to a spring element, which consists of a long flexible segment with a rigid sliding portion at the top. The kinematic pairs in the fixed-guided switch are similar to those found in the fixed-guided switch, with a rolling pair and sliding pair replacing the higher pair at the slider/flexure interface. The locations of the characteristic pivots are $0.85(L/2)$ units from the center of the compliant segment, where L is the total length of that segment. The stiffness of each torsional spring is $K = 2.25 \, EI/(L/2)$. The motion of the switch can be found using the PRBM and standard mechanism analysis (vector loop) equations. The force required to deflect the beam can be found using Newton's laws for the links shown in the PRBM.

Clearly, there is considerable design freedom in the size, material choices and flexure types used in a simple switch of this kind. All else (material, size, etc.) being equal, the switch of case II will have the lowest stress, and lowest actuation force. This is because the motion transferred from the slider to the compliant flexure is distributed over the length of a compliant member, rather than concentrated in a small-length flexural pivot or at the top and bottom of a fixed guided beam.

A good rule of thumb to use is that stresses and forces can be decreased by increasing the length of the flexure and vice versa.

5.4 Conclusions

The pseudo-rigid-body model is a simple, elegant, and easy-to-use way of designing compliant mechanisms. It permits the mechanism problem and the elastic deflection problem to be solved using mechanism techniques. This method is useful in analysis and design because it allows for both computation and intuition about a compliant mechanisms motion.

Acknowledgments

The idea for the compliant switch was suggested by a compliant mechanisms class project (Fall 2009) at the University of South Florida by Glenn Currey, Jantzen Maynard, and Melanie Sutherland.

References

[1] Norton, R. L. *Design of Machinery*, *5th edn*, McGraw-Hill, New York, NY, 2012.

[2] Uicker, J. J., Pennock, G.R., and Shigley, J.E. *Theory of Machines and Mechanisms*, *4th edn*, Oxford, New York, NY, 2011.

[3] Erdman, A.G., Sandor, G. N., and Kota, S. *Mechanism Design: Analysis and Synthesis* Vol I., Prentice-Hall, Upper Saddle River, New Jersey, 2001.

[4] Howell, L. L. *Compliant Mechanisms*, Wiley, New York, NY, 2001.

[5] Lusk, C. P. 2011, "Quantifying Uncertainty for Planar Pseudo-Rigid-Body Models" in *Proceedings of the International Design Engineering Technical Conference IDETC/CIE*, Washington, D.C., USA. Aug. 28–31, 2011.

Appendix: Pseudo-Rigid-Body Examples (by Larry L. Howell)

This section provides examples of pseudo-rigid body models for various loading conditions and beam shapes. Many of these are excerpts from Appendix E of [1].

The pseudo-rigid-body model is used to predict the deflection of large-deflection beams. It is assumed that the flexible part of the beams have a constant cross section, are rigid in shear, have homogeneous material properties, and operate in the elastic range.

A.1.1 Small-Length Flexural Pivot

Description: a flexible segment that is small in length compared to the rigid segments to which it is attached [i.e. $l \ll L$ and $(EI)_l \ll (EI)_L$]. See Figure A.5.1. The characteristic pivot is located at the center of the flexible beam [2].

$$a = \frac{l}{2} + (L + \frac{l}{2})\cos\Theta \qquad b = (L + \frac{l}{2})\sin\Theta \qquad (A.1)$$

$$\theta_o = \frac{M_o l}{EI} \qquad\qquad K = \frac{EI}{l} \qquad (A.2)$$

$$\sigma_{max} = \begin{cases} \dfrac{M_o c}{I} & \text{(loaded with an end moment, } M_o\text{)} \\[3mm] \dfrac{Pac}{I} & \text{(loaded with a vertical force at the free end, } P\text{)} \quad (A.3) \\[3mm] \pm\dfrac{P(a+nb)c}{I} - \dfrac{nP}{A} & \text{(for vertical force, } P\text{, and horizontal force, } nP\text{)} \end{cases}$$

where the maximum stress occurs at the fixed end and c is the distance from the neutral axis to the outer surface of the beam (i.e. half the beam height for rectangular beams, the radius of circular cross section beams, etc.)

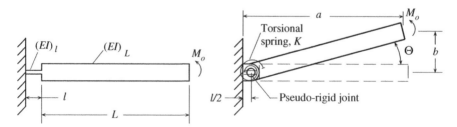

Figure A.5.1   Pseudo-rigid-body model of a small-length flexural pivot

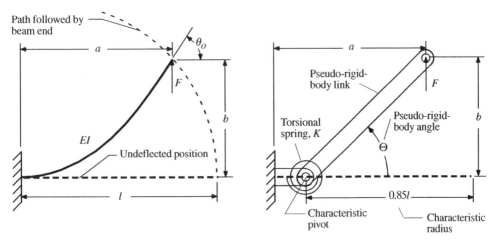

Figure A.5.2 Pseudo-rigid-body model of a vertical force at the free end of a cantilever beam

A.1.2 Vertical Force at the Free End of a Cantilever Beam

Description: a special case of the model of Section 8 that applies to a cantilever beam with a vertical force at the free end ($n = 0$) [3, 4]. See Figure A.5.2.

$$a = l[1 - 0.85(1 - \cos \Theta)] \qquad b = 0.85l \sin \Theta \qquad (A.4)$$

$$\Theta < 64.3 \deg \quad \text{for accurate position prediction} \qquad (A.5)$$

$$\theta_o = 1.24\Theta \qquad K = 2.258\frac{EI}{l} \qquad (A.6)$$

$$\Theta < 58.5 \deg \quad \text{for accurate force prediction} \qquad (A.7)$$

$$F = \frac{K\Theta}{\gamma l \cos \Theta} \qquad (A.8)$$

$$\sigma_{max} = \frac{Pac}{I} \quad \text{at the fixed end} \qquad (A.9)$$

where c is the distance from the neutral axis to the outer surface of the beam (i.e. half the beam height for rectangular beams, the radius of circular cross section beams, etc.)

A.1.3 Cantilever Beam with a Force at the Free End

Description: a beam for which the angle of the force is described by the ratio of the horizontal to vertical components, n. In a compliant mechanism, this represents a flexible beam with a pin joint at one end [3, 4]. See Figure A.5.3.

$$a = l[1 - \gamma(1 - \cos \Theta)] \qquad b = \gamma l \sin \Theta \qquad (A.10)$$

$$\Theta < \Theta_{max}(\gamma) \quad \text{for accurate position prediction} \qquad (A.11)$$

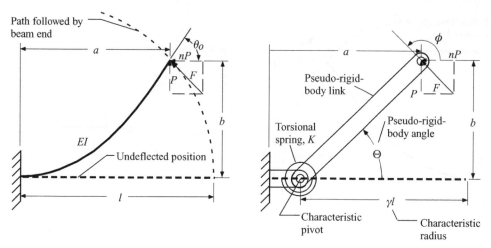

Figure A.5.3 Pseudo-rigid-body model of a cantilever beam with a force at the free end

$$\theta_o = c_\theta \Theta \qquad K = \gamma K_\Theta \frac{EI}{l} \tag{A.12}$$

$$\Theta_{max} < \Theta_{max}(K_\Theta) \quad \text{for accurate force prediction} \tag{A.13}$$

$$\phi = \arctan \frac{1}{-n} \tag{A.14}$$

$$\gamma = \begin{cases} 0.841655 - 0.0067807n + 0.000438n^2 & (0.5 < n < 10.0) \\ 0.852144 - 0.0182867n & (-1.8316 < n < 0.5) \\ 0.912364 + 0.0145928n & (-5 < n < -1.8316) \end{cases} \tag{A.15}$$

$$K_\Theta = \begin{cases} 3.024112 + 0.121290n + 0.003169n^2 & (-5 < n \le -2.5) \\ 1.967647 - 2.616021n - 3.738166n^2 - 2.649437n^3 \\ \quad -0.891906n^4 - 0.113063n^5 & (-2.5 < n \le -1) \\ 2.654855 - 0.509896 \times 10^{-1}n + 0.126749 \times 10^{-1}n^2 \\ \quad -0.142039 \times 10^{-2}n^3 + 0.584525 \times 10^{-4}n^4 & (-1 < n \le 10) \end{cases} \tag{A.16}$$

Or, for a quick approximation: $\gamma = 0.85$ and $K_\Theta = 2.65$.

$$P = \frac{K\Theta}{\gamma l(\cos \Theta + n \sin \Theta)} \qquad \text{or} \qquad F = P\sqrt{1 + n^2} \tag{A.17}$$

$$\sigma_{max} = \pm \frac{P(a + nb)c}{I} - \frac{nP}{A} \quad \text{at the fixed end} \tag{A.18}$$

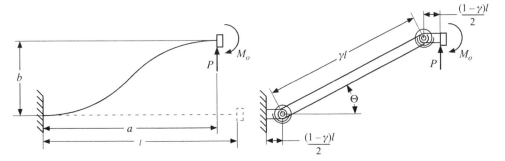

Figure A.5.4   Pseudo-rigid-body model of a fixed-guided beam

where c is the distance from the neutral axis to the outer surface of the beam (i.e. half the beam height for rectangular beams, the radius of circular cross section beams, etc.)

A.1.4 Fixed-Guided Beam

Description: a beam that is fixed at one end; the other end goes through a deflection such that the angular deflection at the end remains constant, and the beam shape is antisymmetric about the center. See Figure A.5.4. This type of beam occurs in parallel-motion mechanisms. The moment, M_o, is a reaction moment required to maintain the constant beam end angle [1, 5].

$$a = l[1 - \gamma(1 - \cos \Theta)] \qquad b = \gamma l \sin \Theta \qquad (A.19)$$

$$\Theta < \Theta_{max}(\gamma) \quad \text{for accurate position prediction} \qquad (A.20)$$

$$\theta_o = 0 \qquad K = 2\gamma K_\Theta \frac{EI}{l} \qquad (A.21)$$

$$\Theta_{max} < \Theta_{max}(K_\Theta) \text{ for accurate force prediction} \qquad (A.22)$$

See Section A1.3 for values of γ and K_Θ.

$$P = \frac{4K_\Theta EI \Theta}{l^2 \cos \Theta} \qquad (A.23)$$

$$\sigma_{max} = \frac{Pac}{2I} \quad \text{at both ends of the beam} \qquad (A.24)$$

where c is the distance from the neutral axis to the outer surface of the beam (i.e. half the beam height for rectangular beams, the radius of circular cross section beams, etc.)

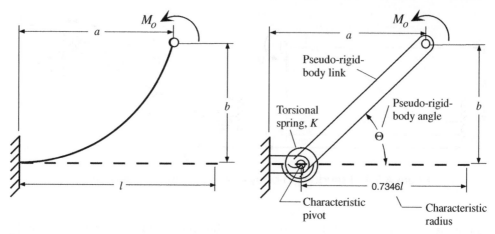

Figure A.5.5 Pseudo-rigid-body model of a cantilever beam with an applied moment at the free end

A.1.5 Cantilever Beam with an Applied Moment at the Free End

Description: a flexible cantilever beam that is loaded with a moment at the free end [1]. See Figure A.5.5.

$$a = l[1 - 0.7346(1 - \cos \Theta)] \qquad b = 0.7346l \sin \Theta \qquad (A.25)$$

$$\Theta_o = 1.5164\Theta \qquad K = 1.5164\frac{EI}{l} \qquad (A.26)$$

$$\sigma_{\max} = \frac{M_o c}{I} \qquad (A.27)$$

where c is the distance from the neutral axis to the outer surface of the beam (i.e. half the beam height for rectangular beams, the radius of circular cross section beams, etc.)

A.1.6 Initially Curved Cantilever Beam

Description: a cantilever beam with an undeflected shape that has a constant radius of curvature, and a force at the free end [6]. See Figure A.5.6.

$$\kappa_o = \frac{l}{R_i} \qquad \Theta_i = \arctan\frac{b_i}{a_i - l(1 - \gamma)} \qquad (A.28)$$

$$\rho = \left[\left(\frac{a_i}{l} - (1 - \gamma)\right)^2 + \left(\frac{b_i}{l}\right)^2\right]^{1/2} \qquad (A.29)$$

$$\frac{a_i}{l} = \frac{1}{\kappa_o}\sin\kappa_o \qquad \frac{b_i}{l} = \frac{1}{\kappa_o}(1 - \cos\kappa_o) \qquad (A.30)$$

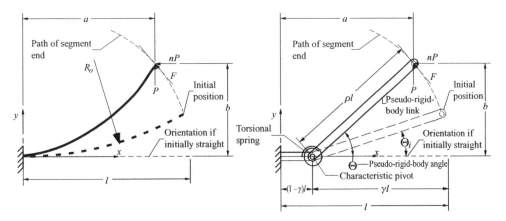

Figure A.5.6 Pseudo-rigid-body model of an initially curved cantilever beam

$$\frac{a}{l} = 1 - \gamma + \rho \cos \Theta \qquad \frac{b}{l} = \rho \sin \Theta \qquad \text{(A.31)}$$

$$K = \rho K_\Theta \frac{EI}{l} \qquad \text{(A.32)}$$

$$\sigma_{max} = \pm \frac{P(a + nb)c}{I} - \frac{nP}{A} \quad \text{at fixed end} \qquad \text{(A.33)}$$

where c is the distance from the neutral axis to the outer surface of the beam (i.e. half the beam height for rectangular beams, the radius of circular cross section beams, etc.)

Table A.5.1 lists values for γ, ρ, and K_Θ for various values of κ_o.

A.1.7 Pinned-Pinned Segments

Description: flexible segments with forces at the ends and no applied moments. See Figure A.5.7. These segments can be modeled as a spring pinned at both ends. The spring constant depends on the geometry and material properties used. The following section provides a model for a common type of pinned-pinned segment.

Table A.5.1 Values for γ, ρ, and K_Θ for various values of κ_o

κ_o	γ	ρ	K_Θ
0.00	0.85	0.850	2.65
0.10	0.84	0.840	2.64
0.25	0.83	0.829	2.56
0.50	0.81	0.807	2.52
1.00	0.81	0.797	2.60
1.50	0.80	0.775	2.80
2.00	0.79	0.749	2.99

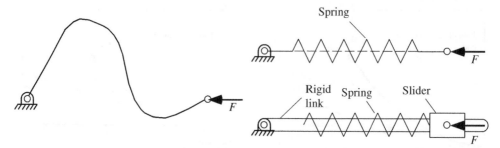

Figure A.5.7 Pseudo-rigid-body model of pinned-pinned segments

Initially Curved Pinned-Pinned Segments

Description: an initially curved beam with an undeflected shape that has a constant radius of curvature, and both ends are pinned [7]. See Figure A.5.8.

Initial coordinates:

$$\frac{a_i}{l} = \frac{1}{\kappa_o}\sin\kappa_o \qquad \frac{b_i}{l} = \frac{1}{2\kappa_o}(1 - \cos\kappa_o) \tag{A.34}$$

$$\kappa_o = \frac{l}{2R_i} \qquad \Theta_i = \arctan\frac{2b_i}{a_i - l(1-\gamma)} \tag{A.35}$$

$$a = l(1 - \gamma + \rho\cos\Theta) \qquad b = \frac{l}{2}\rho\sin\Theta \tag{A.36}$$

$$K = 2\rho K_\Theta\frac{EI}{l} \tag{A.37}$$

$$\rho = \left[\left(\frac{a_i}{l} - (1-\gamma)\right)^2 + \left(\frac{2b_i}{l}\right)^2\right]^{1/2} \tag{A.38}$$

$$\gamma = \begin{cases} 0.8063 - 0.0265\kappa_o & 0.500 \le \kappa_o \le 0.595 \\ 0.8005 - 0.0173\kappa_o & 0.595 \le \kappa_o \le 1.500 \end{cases} \tag{A.39}$$

$$K_\Theta = 2.568 - 0.028\kappa_o + 0.137\kappa_o^2 \quad \text{for } 0.5 \le \kappa_o \le 1.5 \tag{A.40}$$

Table A.5.2 lists values for γ, ρ, K_Θ and $\Delta\Theta_{max}$ for each for various values of κ_o.

$$\sigma_{max} = \pm\frac{Fbc}{I} - \frac{F}{A} \quad \text{at midlength of segment} \tag{A.41}$$

Table A.5.2 Pseudo-rigid-body link characteristics for initially curved pinned-pinned segment

κ_o	γ	ρ	$\Delta\Theta_{max}(\gamma)$	K_Θ	$\Delta\Theta_{max}(K_\Theta)$
0.50	0.793	0.791	1.677	2.59	0.99
0.75	0.787	0.783	1.456	2.62	0.86
1.00	0.783	0.775	1.327	2.68	0.79
1.25	0.779	0.768	1.203	2.75	0.71
1.50	0.775	0.760	1.070	2.83	0.63

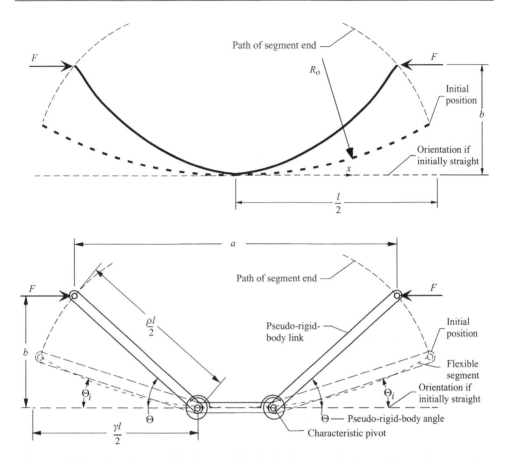

Figure A.5.8 Pseudo-rigid-body model of initially curved pinned-pinned segments

where c is the distance from the neutral axis to the outer surface of the beam (i.e. half the beam height for rectangular beams, the radius of circular cross section beams, etc.)

A.1.8 Combined Force-Moment End Loading

Description: an initially straight flexible segment with a force and moment at the end, such as occurs when both ends are fixed to rigid segments that can move relative to each other. See Figure A.5.9. This approximation is less accurate than the other pseudo-rigid-body models discussed above, but it is presented here as a starting point for problems with flexible segments that have this type of loading condition [8].

$$a = l[1 - \gamma(1 - \cos \Theta)] \qquad b = \gamma l \sin \Theta \qquad (A.42)$$

$$K = 2\gamma K_\Theta \frac{EI}{l} \qquad (A.43)$$

Use values from Section A.5.3 for γ and K_Θ.

Table A.5.3 Values for γ and K_Θ for various loading conditions

Loading Condition	γ_o	γ_1	γ_2	γ_3	K_{Θ_1}	K_{Θ_2}	K_{Θ_3}
General loading (Chen)	0.125	0.351	0.388	0.136	3.25	2.84	2.95
General loading (Su)	0.1	0.35	0.40	0.15	3.51	2.99	2.58
Pure moment	0.1	0.35	0.40	0.15	3.52	2.79	2.80
Pure force	0.1	0.35	0.40	0.15	3.72	2.87	2.26

A.1.9 Combined Force-Moment End Loads – 3R Model

Description: an initially straight flexible segment with a force and moment at the end, such as occurs when both ends are fixed to rigid segments that can move relative to each other. See Figure A.5.10. This pseudo-rigid-body model is more accurate than the model described above with the tradeoff of an increase in complexity. This model was developed by Su [9] with refinements by Chen et al. [10].

$$1 = \gamma_o + \gamma_1 + \gamma_2 + \gamma_3 \tag{A.44}$$

$$\frac{a}{l} = \gamma_o + \gamma_1 \cos\Theta_1 + \gamma_2 \cos(\Theta_1 + \Theta_2) + \gamma_3 \cos(\Theta_1 + \Theta_2 + \Theta_3) \tag{A.45}$$

$$\frac{b}{l} = \gamma_o + \gamma_1 \sin\Theta_1 + \gamma_2 \sin(\Theta_1 + \Theta_2) + \gamma_3 \sin(\Theta_1 + \Theta_2 + \Theta_3) \tag{A.46}$$

$$\theta_o = \Theta_1 + \Theta_2 + \Theta_3 \qquad K_i = K_{\Theta_i}\frac{EI}{l} \tag{A.47}$$

A.1.10 Cross-Axis Flexural Pivot

Description: a cross-axis flexural pivot with a moment load on the end [11]. See Figure A.5.11.

$$K = \frac{K_\Theta EI}{2l} \tag{A.48}$$

$$K_\Theta = 5.300 - 1.687n + 0.885n^2 - 0.209n^3 + 0.018n^4 \tag{A.49}$$

$$K_\Theta = 4.31 \quad \text{for} \quad n = 1(r = w) \text{ where } n = r/w \tag{A.50}$$

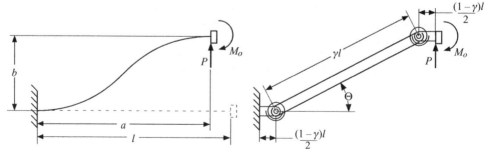

Figure A.5.9 Pseudo-rigid-body model of combined force-moment end loading

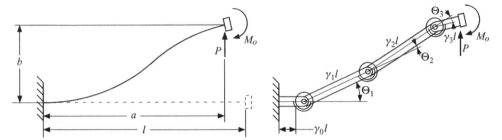

Figure A.5.10 Pseudo-rigid-body 3R model of combined force-moment end loading

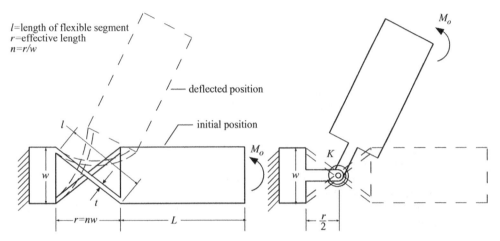

Figure A.5.11 Pseudo-rigid-body model of a cross-axis flexural pivot

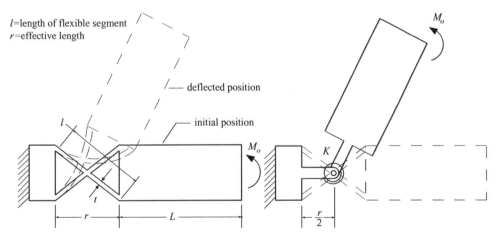

Figure A.5.12 Pseudo-rigid-body model of a cartwheel flexure

A.1.11 Cartwheel Flexure

Description: a cartwheel flexure is similar to a cross-axis flexural pivot except the flexible segments are connected where they cross. Pei et al. [12]. See Figure A.5.12.

$$K = \frac{8EI}{l} \quad \text{for small deflections} \tag{A.51}$$

References

[1] L. L. Howell, *Compliant Mechanisms*. Wiley-Interscience, New York, NY, 2001.

[2] L. L. Howell and A. Midha, "A method for the design of compliant mechanisms with small-length flexural pivots," *Journal of Mechanical Design*, vol. 116, no. 1, pp. 280–290, Mar. 1994.

[3] L. L. Howell, A. Midha, and T. W. Norton, "Evaluation of equivalent spring stiffness for use in a pseudo-rigid-body model of large-deflection compliant mechanisms," *Journal of Mechanical Design*, vol. 118, no. 1, pp. 126–131, 1996.

[4] L. L. Howell and A. Midha, "Parametric deflection approximations for end-loaded, large-deflection beams in compliant mechanisms," *Journal of Mechanical Design*, vol. 117, pp. 156–165, 1995.

[5] C. M. DiBiasio, M. L. Culpepper, R. Panas, L. L. Howell, and S. P. Magleby, "Comparison of molecular simulation and pseudo-rigid-body model predictions for a carbon nanotube–based compliant parallel–guiding mechanism," *Journal of Mechanical Design*, vol. 130, no. 4, pp. 042 308–1 to 042 308–7, 2008.

[6] L. L. Howell and A. Midha, "Parametric deflection approximations for initially curved, large-deflection beams in compliant mechanisms," in *ASME Design Engineering Technical Conferences, DETC96*, 1996.

[7] B. T. Edwards, B. D. Jensen, and L. L. Howell, "A pseudo-rigid-body model for initially-curved pinned-pinned segments used in compliant mechanisms," *Journal of Mechanical Design*, vol. 123, no. 3, pp. 464–468, 2001.

[8] S. M. Lyon and L. L. Howell, "A simplified pseudo-rigid-body model for fixed-fixed flexible segments," in *ASME Design Engineering Technical Conferences, DETC02/DAC*, 2002.

[9] H. Su, "A pseudorigid-body 3r model for determining large deflection of cantilever beams subject to tip loads," *Journal of Mechanisms and Robotics*, vol. 1, no. 2, 2009.

[10] G. Chen, B. Xiong, and X. Huang, "Finding the optimal characteristic parameters for 3r pseudo-rigid-body model using an improved particle swarm optimizer," *Precision Engineering*, vol. 35, no. 3, pp. 505–511, 2011.

[11] B. D. Jensen and L. L. Howell, "The modeling of cross-axis flexural pivots," *Mechanism and Machine Theory*, vol. 37, no. 5, pp. 461–476, 2002.

[12] X. Pei, J. Yu, G. Zong, S. Bi, and H. Su, "The modeling of cartwheel flexural hinges," *Mechanism and Machine Theory*, vol. 44, no. 10, pp. 1900–1909, 2009.

Part Three
Synthesis of Compliant Mechanisms

6

Synthesis through Freedom and Constraint Topologies

Jonathan Hopkins

University of California, Los Angeles, USA

6.1 Introduction

Determining how best to use compliant elements to constrain a rigid body such that it possesses a desired set of degrees of freedom (DOFs) is a difficult but important challenge for compliant mechanism designers. This chapter introduces a specialized synthesis approach called freedom and constraint topologies (FACT) [1–11] that provides a systematic framework and process for designers. The basis for the approach is a comprehensive library of geometric shapes shown in Figure 6.1 that represent the mathematics of screw theory and enable designers to visualize the regions wherein all the compliant constraint elements could be placed that would permit the mechanism's desired DOFs. In this way, designers may rapidly consider every concept that best satisfies the kinematic, elastomechanic, and dynamic design requirements before settling on the final design concept. These shapes contain all the relevant quantitative information that is needed to rapidly generate complex compliant concepts without undue complications that arise when one focuses on detailed mathematical treatments that are better suited for optimization rather than visualization and synthesis. As such, the FACT synthesis process significantly impacts the design of precision motion stages, general purpose flexure bearings, nanomanufacturing equipment, optical manipulation stages, and precision instruments used for nanoscale research.

The FACT library shown in Figure 6.1 contains two sets of complementary geometric shapes that help designers synthesize flexure-based compliant mechanisms like those shown in Figure 6.2. One set of shapes, called freedom spaces, represent the permissible motions or DOFs of a flexure system, and the other set of shapes, called

---

Handbook of Compliant Mechanisms, First Edition. Edited by Larry L. Howell, Spencer P. Magleby and Brian M. Olsen.
© 2013 John Wiley & Sons, Ltd. Published 2013 by John Wiley & Sons, Ltd.

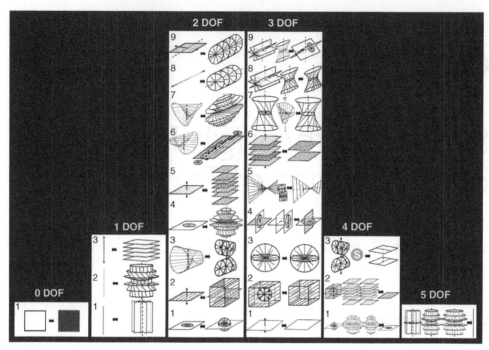

Figure 6.1 FACT library of freedom and constraint spaces for synthesizing a flexure system's topology

constraint spaces, represent the regions wherein the compliant constraints could be placed that would permit those DOFs. Consider, for example, the complementary freedom and constraint spaces labeled 1 in the 1 DOF column of Figure 6.1. A larger depiction of these shapes is provided in Figure 6.2A. The freedom space on the left side of the double-sided arrow is a line that represents a single rotation about its axis. The constraint space on the right side of the double-sided arrow is every plane that intersects this line's axis and represents the regions wherein compliant constraints could be placed for permitting the rotational DOF of the complementary freedom space. The reason that the flexure system shown in Figure 6.2B possesses a single rotational DOF is that the system is constrained with flexure blades that lie on the intersecting planes of the complementary constraint space shown in Figure 6.2C. Similarly, the reason that the flexure system shown in Figure 6.2D possesses the same rotational DOF is that the system is also constrained with flexure blades that lie on the intersecting planes of the complementary constraint space shown in Figure 6.2E. With an understanding of these intersecting planes, therefore, designers may rapidly visualize, generate, and compare every flexure system concept that possesses a single rotational DOF. This idea of comprehensive concept generation using the geometric shapes of Figure 6.1 to achieve any desired set of DOFs is integral to the FACT synthesis process.

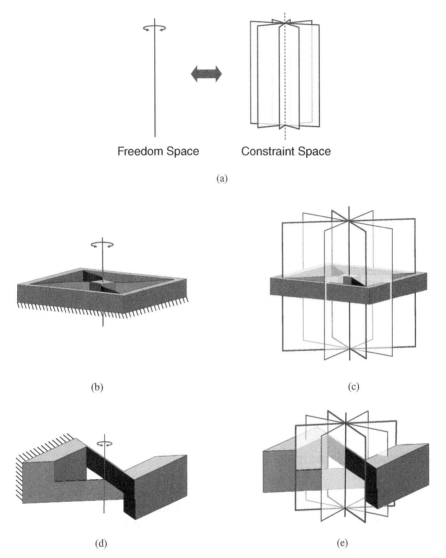

Freedom Space   Constraint Space

(a)

(b)   (c)

(d)   (e)

Figure 6.2 Complementary freedom and constraint space pair (A) for every parallel flexure system that possesses a single rotational DOF. Multiple concepts that possess the motion of the freedom space, (B) and (D), may be generated using the shape of the constraint space, (C) and (E)

Although FACT may be used to synthesize most types of compliant mechanisms, this chapter focuses largely on synthesis of parallel flexure systems. Parallel flexure systems, like those shown in Figure 6.2 and Figure 6.3A, consist of a single rigid stage connected directly to ground by compliant constraints. Serial and hybrid flexure systems, like those shown in Figures 6.3B and C, consist of multiple parallel flexure modules that are nested or stacked together. Although this chapter will enable

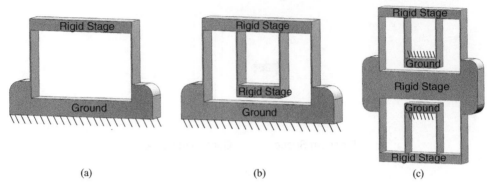

Figure 6.3   Parallel (A), serial (B), and hybrid (C) flexure systems

designers to replace compliant constraints from parallel flexure systems with kine-matically equivalent serial flexure chains, a complete explanation of how FACT may be used to synthesize serial and hybrid flexure systems is beyond the scope of this chapter. Furthermore, the flexure systems synthesized in this chapter are best suited for precision applications (i.e. the system motions are at least three or more orders of magnitude smaller than the size of the mechanism). This chapter does not, therefore, discuss how FACT could be applied to the synthesis of flexure systems that guide stages along desired motion paths, but rather systems that possess desired DOFs or directions of greatest compliance. Lastly, this chapter focuses on kinematic flexure synthesis only. Considerations of stiffness and dynamics are beyond the scope of this chapter.

6.2 Fundamental Principles

This section introduces the underlying principles necessary to understand the FACT synthesis approach. Motion and constraint systems are modeled using screw theory and represented using geometric shapes. The relationship between these shapes is established and discussed in the context of the comprehensive library of shapes from Figure 6.1. Principles of kinematic equivalence are introduced as a means of generating design concept alternatives that possess identical DOFs.

6.2.1 Modeling Motions using Screw Theory

According to screw theory [12–19], any infinitesimal motion is a screw motion that may be modeled as a 6×1 vector called a twist, **T**, and represented by a line along and about which a stage may simultaneously translate and rotate. The ratio of the stage's translation to its rotation is called the pitch of the screw motion. If the pitch is zero, the motion is a rotation. If the pitch is infinite, the motion is a translation. If the pitch is any other value, the motion is a screw. Examples of parallel flexure systems that possess one of these three fundamental motion types are shown in Figure 6.4A–C.

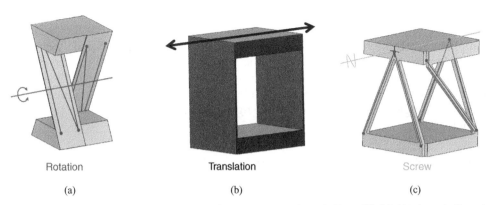

Figure 6.4 Parallel flexure systems that possess a single rotational DOF (A), translational DOF (B), and screw DOF (C)

The parallel flexure systems shown in Figure 6.4 each only possess a single DOF. Consider instead the three DOF parallel flexure system shown in Figure 6.5. The four blade flexures constrain the rigid stage such that it possesses two rotational DOFs shown in Figures 6.5A and B and one translational DOF shown in Figure 6.5C. Although these three motions represent the system's DOFs, they do not represent all the motions permitted by the four blade flexures. If, for instance, all three DOFs were simultaneously actuated with various magnitudes, the stage would appear to rotate about lines that lie on the plane of the blade flexures. This plane of rotation lines and the orthogonal translation arrow shown in Figure 6.5D is the system's freedom

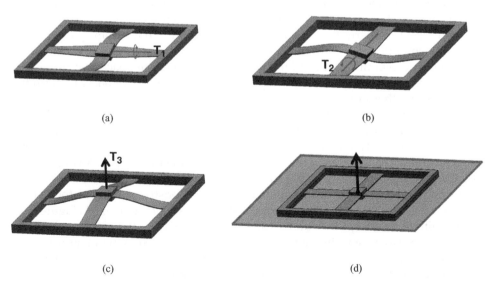

Figure 6.5 A parallel flexure system with three DOFs – two rotations, (A) and (B), and one translation, (C). The freedom space of the parallel flexure system (D)

space. Freedom space is the geometric shape that visually represents the complete kinematics of a constraint system (i.e. all the motions or twists that the system's compliant constraints permit). A system's freedom space may be modeled by linearly combining the twists of its DOFs. For the system shown in Figure 6.5 the planar freedom space would be generated by linearly combining the three DOF twists, T_1, T_2, and T_3. The freedom spaces of the parallel flexure systems shown in Figure 6.4 are simply the single twist lines shown in the figure because each system only possesses a single DOF.

6.2.2 Modeling Constraints using Screw Theory

It is not enough to model motions alone to establish the FACT synthesis approach. Compliant constraints must also be modeled using screw theory. Compliant constraints are only capable of imparting combinations of forces on the stages that they constrain. They may, therefore, be represented by sets of lines that are collinear with the axes of the forces that the constraints are capable of imparting. According to screw theory, each of these lines, called constraint lines, may be modeled using a pure force 6×1 vector called a wrench [12–19], **W**. If a compliant constraint is long and slender, like the wire flexures shown in Figure 6.4C, a single pure force wrench oriented along the constraint's axis correctly models the constraint. If the compliant constraint is a thin blade flexure, like those shown in Figures 6.4A and B, the set of constraint lines that lie on the plane of the blade and directly connect the stage to the ground, accurately model the compliant constraint. Example sets of three constraint lines are drawn on each of the blade flexures shown in Figures 6.4A and B.

The relationship between a flexure system's DOFs and its compliant constraints is embodied by the rule of complementary topologies [1, 2]. This rule states that every freedom space uniquely links to a complementary constraint space. Constraint space is the geometric shape that visually represents the region wherein all the compliant constraints exist for permitting the desired DOFs within the freedom space. From a synthesis stand point, the concept of constraint space is very powerful. If a designer knows which constraint space uniquely links to the freedom space that represents the desired DOFs, he/she is able to very rapidly visualize every concept within the constraint space that satisfies the desired kinematics.

Consider the complementary freedom and constraint space pair shown in Figure 6.6A. Recall that this freedom space is the three DOF flexure system's freedom space from Figure 6.5D. Its complementary constraint space is a plane that represents every constraint line that lies on the same plane as the freedom space. Any constraint system with constraint lines that lie only on this plane will permit the motions within the freedom space. Note from Figure 6.6B that the constraint lines of the stage's four blade flexures lay within the planar constraint space of Figure 6.6A. Figure 6.6C shows a different concept that utilizes six wire flexures from the plane of the same constraint space to achieve the same kinematics as the stage shown in Figure 6.6B.

Although constraints selected from within a system's constraint space will always permit the desired DOFs represented by its complementary freedom space, the correct number of independent constraint lines must be selected to assure that the system doesn't possess extra DOFs as well. If only a single wire flexure had been selected from

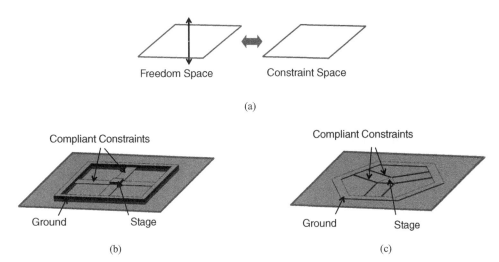

Freedom Space   Constraint Space

(a)

Compliant Constraints   Compliant Constraints

Ground   Stage   Ground   Stage

(b)   (c)

Figure 6.6 Freedom and constraint space pair for a system that possesses the DOFs from Figure 6.5 (A). Multiple flexure system concepts may be generated using the planar constraint space (B) and (C)

within the plane of the constraint space of Figure 6.6A, for instance, the stage would not only possess the DOFs within the desired freedom space, but it would also possess other unwanted DOFs. For a constraint system to only possess the desired n DOFs represented by its freedom space, $6-n$ independent constraint lines must be selected from the freedom space's complementary constraint space. To identify how many constraint lines are independent from among a select group, Gaussian elimination may be applied to the wrench vectors that model them. As an alternative to Gaussian elimination, a comprehensive list of qualitative "rules of thumb" exist for guiding designers in selecting independent constraint lines from within any constraint space. These rules are embodied by shapes called subconstraint spaces and are found in [2, 7]. For the constraint space of Figure 6.6A, at least three constraint lines that are not all parallel and do not all intersect at the same point must be selected. The constraint lines that model the compliant constraint elements shown in Figures 6.6B and C satisfy this condition for both flexure systems.

Once the appropriate number of independent constraint lines has been selected from a constraint space, any other constraint line selected from the same space is redundant and will not affect the system's kinematics. Any three wire flexures from the system shown in Figure 6.6C are examples of redundant constraints. If any three of the wire flexures were removed from the system, the system's DOFs would not change. Although redundant constraints do not affect the system's DOFs, they do affect the system's stiffness, load capacity, dynamics, and symmetry. Constraint space is, therefore, not only important for helping designers synthesizing constraint systems that achieve any desired set of DOFs, but it is also important for helping designers visualize the regions wherein every redundant constraint exists for optimizing other design parameters without affecting the system's desired kinematics.

6.2.3 Comprehensive Library of Freedom and Constraint Spaces

The FACT synthesis approach enables designers to rapidly visualize and consider every compliant constraint topology that enables any desired set of DOFs. The comprehensive nature of this approach is due to the fact that there are a finite number of complementary freedom and constraint space pairs. These pairs or types are shown in Figure 6.1 but are described in detail and derived in Hopkins [6, 7]. For this chapter, the reader is not expected to understand all the information contained in this figure. What is important to understand is that all of the spaces belong to one of six columns where each column pertains to the number of DOFs represented by the freedom spaces within each column. There is no 6-DOF column in the figure because a system that possesses six DOFs is not constrained. Each freedom space is shown to the left of a small, gray, double-sided arrow and its complementary constraint space is shown to the right of the same arrow similar to the freedom and constraint space pair shown in Figure 6.6A. Note that this pair is shown in Figure 6.1 as Type 1 in the 3 DOF column. Furthermore, note that the 1 DOF column contains only three types of freedom and constraint space pairs because only three types of motions exist—translations, screws, and rotations.

It is also important to note that the library of spaces shown in Figure 6.1 is comprehensive for parallel flexure systems only. There are other freedom spaces not shown in Figure 6.1 that are achievable by stacking parallel flexure system modules in series. The complete list of these additional freedom spaces is provided in [6, 20]. Using this list and the library from Figure 6.1 designers are able to rapidly visualize every flexure system (i.e. parallel, serial, and hybrid) that possesses any desired set of DOFs.

6.2.4 Kinematic Equivalence

Compliant constraint elements that possess the same kinematic characteristics but possess different geometric, dynamic, and elastomechanic characteristics are said to be kinematically equivalent [6, 20]. Such compliant constraint elements may be interchanged without altering the flexure system's DOFs. This observation enables designers to consider a multiplicity of other concepts that achieve the same desired kinematic design requirements.

Consider the compliant constraint elements shown in Figure 6.7A. Both the wire flexure constraint and the stacked flexure blade constraint possess the same five DOFs—three orthogonal intersecting rotational DOFs and two orthogonal translational DOFs that are perpendicular to the axis of the wire flexure. These two compliant constraints are kinematically equivalent because they constrain the same directions of motion while permitting the same DOFs. Any design, therefore, that uses a wire flexure to constraint a stage that possesses a desired set of DOFs may be replaced by the stacked blade flexure constraint shown in Figure 6.7A. This observation is powerful because it enables designers to consider other constraint topologies that permit the same kinematics while allowing for a larger variety of buckling, dynamic, and stiffness characteristics. The wire flexure shown in Figure 6.7B is also kinematically equivalent to the bent flexure blade shown on the right side of the figure. It is

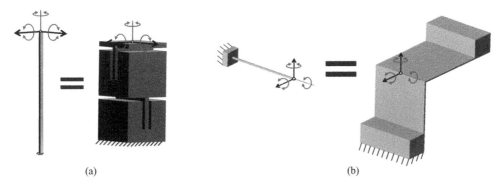

Figure 6.7 Examples of kinematically equivalent compliant constraint elements

important to note, however, that the crease of the bent flexure blade must align with the axis of the wire flexure if it is to impose the same constraint kinematics.

6.3 FACT Synthesis Process and Case Studies

The four steps of the FACT synthesis process for designing parallel flexure systems are as follows:

Step 1: *Identify the desired DOFs.*
Step 2: *Identify the correct freedom space that contains the DOFs from Step 1.*
Step 3: *Select enough nonredundant constraints from the complementary constraint space of the freedom space from Step 2.*
Step 4: (Optional) *Select redundant constraints from the system's constraint space to achieve greater stiffness, load capacity, and symmetry.*

In this section these steps are discussed in detail and applied to the synthesis of two flexure system case studies.

6.3.1 Flexure-Based Ball Joint Probe

The first case study is the design of a probe that is constrained by compliant elements that mimic the kinematics of a ball joint. For step 1 of the FACT design process, therefore, three intersecting and independent rotational DOFs are selected and shown intersecting at the tip of the probe in Figure 6.8A. For step 2, the freedom space that contains these three intersecting rotation lines is the sphere of intersecting rotation lines shown as Type 3 in the 3 DOF column of Figure 6.1. This correct freedom space is shown again with its complementary constraint space in Figure 6.8B. The constraint space is a sphere of intersecting constraint lines. For step 3, at least three independent constraint lines must be selected from within this sphere. To assure independence of these three lines, the lines must not all lie on a common plane. The three wire flexures shown in Figure 6.8C form a tripod configuration with constraint lines that satisfy this condition. Step 4 is unnecessary for this case study unless the designer wishes to

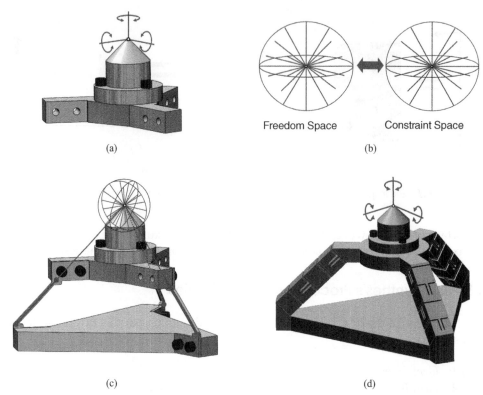

Figure 6.8 Desired motions (A), correct freedom and constraint space pair (B), selecting constraints from the constraint space (C), and replacing the wire flexures with kinematically equivalent compliant constraints (D)

add redundant wire flexures to increase system stiffness and load capacity. Another way to increase the system's stiffness and load capacity without adding redundant constraints, while maintaining the system's intended DOFs, would be to replace the existing wire flexures from Figure 6.8C with kinematically equivalent stacked blade flexures from Figure 6.7A as shown in Figure 6.8D.

6.3.2 X-Y-ThetaZ *Nanopositioner*

The second case study is the design of a nanopositioner that possesses three DOFs— one rotational DOF and two translational DOFs that are perpendicular to the axis of rotation. For step 1 of the FACT design process, therefore, these DOFs are selected and shown on the nanopositioner's stage in Figure 6.9A. For step 2, the freedom space that contains the rotational DOF and the two translational DOFs is the box of parallel rotation lines and the disk of translation arrows that are orthogonal to the rotation lines shown as Type 2 in the 3-DOF column of Figure 6.1. This correct freedom space is shown again with its complementary constraint space in Figure 6.9B. The constraint space is a box of parallel constraint lines that are parallel to the rotation

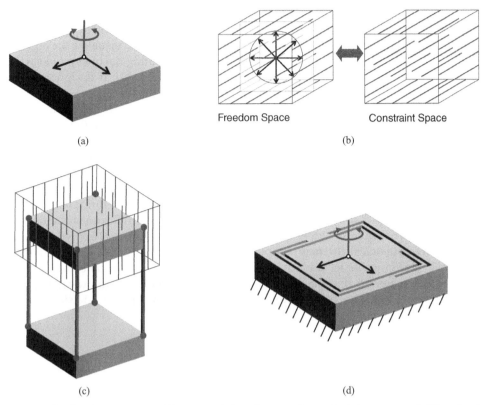

Freedom Space   Constraint Space

(a)   (b)

(c)   (d)

Figure 6.9 Desired motions (A), correct freedom and constraint space pair (B), select-ing constraints from the constraint space (C), and replacing the wire flexures with kinematically equivalent compliant constraints (D)

lines of the freedom space. For step 3, at least three independent constraint lines must be selected from within this box. To assure independence of these three lines, the lines must not all lie on a common plane. The constraint lines of three of the wire flexures shown in Figure 6.9C satisfy this condition. For step 4 a fourth wire flexure is selected from the constraint space to improve the system's symmetry. If the four wire flexures shown in Figure 6.9C were replaced with bent flexure blades from Figure 6.7B, the system would possess the same desired DOFs but it would also possess better dynamic characteristics. Furthermore, this design would be more compact and easier to fabricate. It would also be impervious to actuated parasitic errors as well as fluctuations in temperature.

6.4 Current and Future Extensions of FACT's Capabilities

This chapter has briefly touched upon the capabilities of the FACT synthesis process. If designers wish to extend these principles to the synthesis of serial and hybrid flexure systems see Hopkins and coworkers [6, 20, 21]. If designers wish to synthesize parallel flexure systems that mimic the complex kinematics represented by freedom spaces that are only achievable using serial and hybrid flexure systems see Hopkins [22]. If designers wish to use the geometric shapes of the FACT library to visualize the regions wherein designers could optimally place actuators for actuating flexure systems with minimal parasitic errors see Hopkins and coworkers [6, 23, 24]. If designers wish to extend these principles to the synthesis of flexure systems that possess geometric or mechanical advantages with specified transmission ratios see Hopkins and Panas [25, 26]. If designers wish to utilize the principles of FACT to analyze the sensitivity of compliant constraint geometry and orientation see Dibiasio and Hopkins [27]. If designers wish to understand the geometry of the FACT shapes in greater detail see Hopkins [6, 7].

Research is currently underway that will extend the FACT synthesis capabilities further. Future capabilities will include the synthesis of (i) large deformation flexure systems with stages that move along specified motion paths, (ii) flexure systems that possess desired dynamic characteristics (e.g., particular mode shapes excited at specified natural frequencies), (iii) new compliant constraint elements that achieve unusual kinematic, elastomechanic, and dynamic characteristics, (iv) flexible microstructures for new materials that possess naturally unobtainable physical properties, and (v) new bearing types for both compliant and rigid system's that guide stages with complex DOFs.

Acknowledgments

This chapter was performed under the auspices of the U.S. Department of Energy by Lawrence Livermore National Laboratory under Contract DE-AC52-07NA27344. LLNL-BOOK-506451.

References

[1] Hopkins, J.B. and Culpepper, M.L., "Synthesis of multi-degree of freedom, parallel flexure system concepts via freedom and constraint topology (FACT)—Part I: Principles," *Precision Engineering*, **34** (2010): pp. 259–270.

[2] Hopkins, J.B. and Culpepper, M.L., "Synthesis of multi-degree of Freedom, Parallel Flexure System Concepts via Freedom and Constraint Topology (FACT)—Part II: Practice," *Precision Engineering*, **34** (2010): pp. 271–278.

[3] Hopkins, J.B. and Culpepper, M.L., "A Design Process for the Creation of Precision Flexure Concepts via the Use of Freedom and Constraint Topology," *Proc. of the Annual Meeting of the American Society for Precision Engineering*, Dallas, TX, October 2007.

[4] Hopkins, J.B. and Culpepper, M.L., "Synthesis of Multi-Degree of Freedom Precision Flexure Concepts via Freedom and Constraint Topologies," *Proc. of the Annual Meeting of the American Society for Precision Engineering*, Dallas, TX, October 2007.

[5] Hopkins, J.B. and Culpepper, M.L., "A Quantitative, Constraint-based Design Method for Multi-axis Flexure Stages for Precision Positioning and Equipment," *Proc. of the Annual Meeting of the American Society for Precision Engineering*, Monterey, CA, October 2006, pp. 139–42.

[6] Hopkins, J.B., Design of flexure-based motion stages for mechatronic systems via free-dom, actuation and constraint topologies (FACT). PhD Thesis. Massachusetts Institute of Technology; 2010.

[7] Hopkins, J.B., Design of parallel flexure systems via freedom and constraint topologies (FACT). Masters Thesis. Massachusetts Institute of Technology; 2007.

[8] Yu, J.J., Kong, X.W., Hopkins, J.B., Culpepper, M.L., and Dai., J.S., "The Reciprocity of a Pair of Line Spaces," *Proc. of the 13th World Congress in Mechanism and Machine Science,* Guanajuato, México, June, 2011.

[9] Yu, J.J., Li, S.Z., Pei, X., Su, H., Hopkins, J.B., and Culpepper, M.L., "Type Synthesis Prin-ciple and Practice of Flexure Systems in the Framework of Screw Theory Part I: General Methodology," *Proc. of the ASME 2010 International Design Engineering Technical Confer-ences & Computers and Information in Engineering Conference IDETC/CIE 2010,* Montreal, Quebec, Canada, August 2010.

[10] Yu, J.J., Li, S.Z., Pei, X., Su, H., Hopkins, J.B., and Culpepper, M.L., "Type Synthesis Principle and Practice of Flexure Systems in the Framework of Screw Theory Part II: Numerations and Synthesis of Large-Displacement Flexible Joints," *Proc. of the ASME 2010 International Design Engineering Technical Conferences & Computers and Information in Engineering Conference IDETC/CIE 2010,* Montreal, Quebec, Canada, August 2010.

[11] Yu, J.J., Li, S.Z., Pei, X., Su, H., Hopkins, J.B., and Culpepper, M.L., "Type Synthesis Principle and Practice of Flexure Systems in the Framework of Screw Theory Part III: Numerations and Type Synthesis of Flexure Mechanisms," *Proc. of the ASME 2010 Interna-tional Design Engineering Technical Conferences & Computers and Information in Engineering Conference IDETC/CIE 2010,* Montreal, Quebec, Canada, August 2010.

[12] Ball, R.S., *A Treatise on the Theory of Screws.* Cambridge, UK: The University Press; 1900.

[13] Phillips, J., *Freedom in Machinery: Volume 1, Introducing Screw Theory.* New York, NY: Cambridge University Press; 1984.

[14] Phillips, J., *Freedom in Machinery: Volume 2, Screw Theory Exemplified.* New York, NY: Cambridge University Press; 1990.

[15] Bothema, R. and Roth, B., *Theoretical Kinematics.* Dover, New York, 1990.

[16] Hunt, K.H., *Kinematic Geometry of Mechanisms.* Oxford, UK: Clarendon Press; 1978.

[17] Merlet, J.P., Singular configurations of parallel manipulators and Grassmann geometry. *Inter J of Robotics Research* 1989;8(5):45–56.

[18] Hao, F. and McCarthy, J.M., Conditions for line-based singularities in spatial platform manipulators. *J of Robotic Sys* 1998;15(1):43–55.

[19] McCarthy, J.M., *Geometric Design of Linkages.* Cambridge, MA: Springer Press; 2000.

[20] Hopkins, J.B. and Culpepper, M.L., "Synthesis of Precision Serial Flexure Systems Using Freedom and Constraint Topologies (FACT)," *Precision Engineering,* **35** (2011): pp. 638–649.

[21] Hopkins, J.B. and Culpepper, M.L., "Synthesis of Multi-Axis Serial Flexure Systems," *Proc. of the Annual Meeting of the American Society for Precision Engineering,* Monterey, CA, October 2009, pp. 116–119.

[22] Hopkins, J.B., "Synthesizing Parallel Flexure Concepts that Mimic the Complex Kinemat-ics of Serial Flexures Using Displaced Screw Systems," *Proc. of the ASME 2011 International Design Engineering Technical Conferences & Computers and Information in Engineering Con-ference IDETC/CIE 2011,* Washington, DC, USA, August 2011.

[23] Hopkins, J.B. and Culpepper, M.L., "A Screw Theory Basis for Quantitative and Graphical Design Tools that Define Layout of Actuators to Minimize Parasitic Errors in Parallel Flexure Systems," *Precision Engineering,* **34** (2010): pp. 767–776.

[24] Hopkins, J.B. and Culpepper, M.L., "Determining the Optimal Actuator Placement for Parallel Flexure Systems," *Proc. of the 9th International Conference of the European Society for Precision Engineering & Nanotechnology,* San Sebastian, Spain, June 2009.

[25] Hopkins, J.B. and Panas, R.M., "Design of Flexure-based Precision Transmission Mechanisms using Screw Theory," submitted to *Precision Engineering*, June 2011.

[26] Hopkins, J.B. and Panas, R.M., "Design of Flexure-based Precision Transmission Mechanisms Using Screw Theory," *Proc of the 11th International Conference of the European Society for Precision Engineering & Nanotechnology*, Como, Italy, May 2011.

[27] DiBiasio, C.M. and Hopkins, J.B., 2012, "Sensitivity of Freedom Spaces During Flexure Stage Design via FACT," *Precision Engineering*, **36**(3): pp. 494–499.

7

Synthesis through Topology Optimization

Mary Frecker

The Pennsylvania State University, USA

This chapter describes an approach for synthesizing compliant mechanisms that uses topology optimization to meet particular functional needs. Topology optimization techniques are especially useful when the designer does not have a particular compliant mechanism already in mind. This approach can also be used to augment intuition-based or experience-based compliant mechanism designs. Topology optimization can result in novel solutions that the designer might not have arrived at by means such as converting a known rigid-link mechanism to a compliant mechanism. It is intended to predict the best topology, or material connectivity in a compliant structure, for a particular compliant mechanism design problem. Topology optimization is widely used in a variety of structural design problems; the discussion here is focused on topology synthesis of compliant mechanisms.

7.1 What is Topology Optimization?

Topology is defined as the pattern of connectivity or spatial sequence of members or elements in a structure. The allowable space for the design in a topology optimization problem is called the design domain. The topology is defined by the distribution of material and void within the design domain (Figure 7.1). Nondesign elements (solid or void) can be specified and are not changed by the optimizer. For example, the designer may require that a certain portion of the design domain remain empty; this region would be specified as void nondesign.

Handbook of Compliant Mechanisms, First Edition. Edited by Larry L. Howell, Spencer P. Magleby and Brian M. Olsen.
© 2013 John Wiley & Sons, Ltd. Published 2013 by John Wiley & Sons, Ltd.

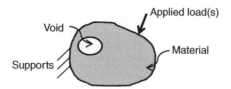

Figure 7.1 The design domain is the allowable space for the design and consists of regions of material and void. Supports and loads are also specified

The example pictured in Figure 7.2 illustrates the differences between topology, geometry, and size optimization. A rectangular design domain is pictured with supports on the left hand side and a downward load at the lower right corner (Figure 7.2a). An initial ground structure (described in more detail in Section 7.3) is used in a classical structural design problem where the goal is to minimize structural compliance and weight. This problem can be thought of as the design of the stiffest, least-weight structure. The optimal topology consists of a subset of elements from the initial ground structure, as pictured in Figure 7.2b. This topology is then refined using geometry optimization (Figure 7.2c), where essentially the locations of the nodes that connect elements are adjusted to improve the compliance and weight. The cross-sectional size of the elements themselves can then be adjusted using size optimization to further improve the compliance and weight (Figure 7.2d). The structures pictured in

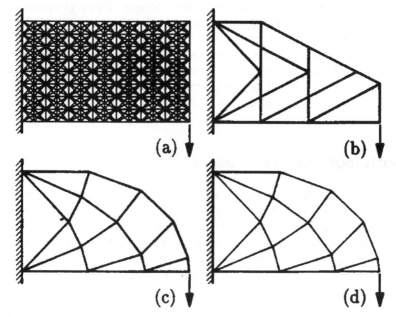

Figure 7.2 (a) Ground structure, (b) optimal topology solution for minimum structural compliance and weight, (c) structure improved by geometry optimization, (d) further improvement by size optimization (adapted from (1))

Figures 7.2b–d all have the same topology; Figures 7.2b and c have the same topology but different geometry, and Figures 7.2c and d have the same geometry but different sizes.

An important quantity used in topology optimization problems is the volume fraction F_v (Eq. (7.1)), which is the ratio of the volume occupied by solid material (V_M) to the total available volume in the design domain (V_A). Typically topology optimization problems are formulated with an upper limit constraint on the volume fraction. This quantity is sometimes referred to as the material resource constraint.

$$F_v = \frac{V_M}{V_A} \qquad (7.1)$$

7.2 Topology Optimization of Compliant Mechanisms

In the synthesis of compliant mechanisms, topology optimization is used to design a flexible structure with a specified output displacement in response to the input force. The displacement inverter shown in Figure 7.3a is an example that is commonly

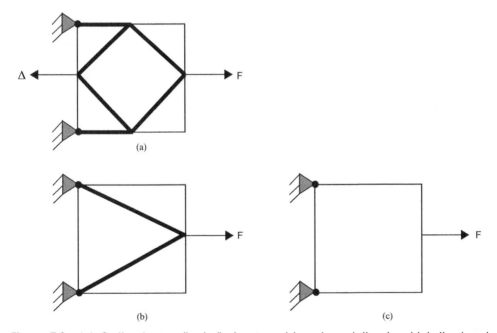

Figure 7.3 (a) Optimal compliant displacement inverter solution in which the input force F and output displacement Δ are in opposite directions; (b) Minimum compliance solution is the stiffest possible structure where there is very little deformation in response to the input force; (c) Maximum compliance solution consists of no material

used in the development of compliant mechanism topology design problems. In this example, the design domain is indicated by the outer black square, and the supports fix the horizontal and vertical degrees of freedom at the upper and lower corners of the left edge. Here, the optimal compliant mechanism topology is defined by the bold black lines. The elements in this compliant mechanism can be connected by pin joints or solid joints with no rotational degrees of freedom; the choice of elements is discussed further in Sections 7.3 and 7.4. In either case, the output displacement Δ is in the opposite direction as the input force F, hence the name "displacement inverter".

Contrary to the classic structural design problem where the stiffest possible structure is desired, the compliant mechanism design problem is a compromise between flexibility and stiffness. The stiffest possible structure, or minimum compliance solution, will exhibit very little output displacement in response to the input load. For the same design domain, supports, and applied load, the minimum compliance solution is pictured in Figure 7.3b. One might think that for compliant mechanism design we should simply *maximize*, rather than minimize, compliance. However, if we consider the most flexible possible structure, or maximum compliance, the best solution is actually one with no material, i.e. no structure at all (Figure 7.3c). Clearly this solution is not practical. In fact, the best compliant mechanism design is a compromise between these two problems.

This compromise can be further elucidated through the following example. Consider the common u-shaped plastic salad bar tongs, which are in fact a compliant mechanism. If the tongs are too flexible they cannot effectively grasp. That is, it is easy to deform the tongs, but little of the input energy is actually available to grasp the salad. On the other hand, if the tongs are too stiff, they will not deform enough to grasp the salad. So the best tongs are actually a compromise between flexibility and stiffness. An analogous example is one in which we would like to design a compliant mechanism to act as a gripper. The compliant gripper must be flexible enough so that when the loads are applied, it easily deforms and closes around the workpiece. However, if the compliant gripper is too flexible it will not be able to apply sufficient force to the workpiece, i.e. much of the input energy will be expended in deforming the very flexible gripper and little will be transferred to the workpiece. So, the compliant gripper must be stiff enough to apply sufficient force to grip and hold the workpiece, but not so stiff that the workpiece is out of reach. We use topology optimization in compliant mechanism design to effectively resolve this trade-off between flexibility and stiffness.

There have been a number of formulations developed to handle the trade-off between flexibility and stiffness in compliant mechanism design; a review and a comparison of these formulations is presented in [2]. One of the original formulations [3] is summarized here. The design problem is broken down into two parts, as pictured in Figures 7.4a and 4b. Figure 7.4a shows a generic design domain with prescribed supports and an applied load F_A located at the input point A. In this condition we would like the compliant mechanism to be flexible so that it will easily deform in the direction of the desired output deflection in response to the applied load. A virtual or "dummy" load F_B is applied at the output point B in the direction of the desired output deflection Δ. A second loading condition is also considered where

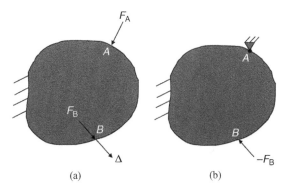

(a)   (b)

Figure 7.4 Compliant mechanism design problem. (a) F_A is the applied force at input point A, Δ is the desired output deflection at output point B, and F_B is a virtual load at point B in the direction of Δ; (b) The virtual load is applied in the opposite direction to represent the resistance of the workpiece

the virtual load is applied at point B in the opposite direction (Figure 7.4b), and can be thought of as the resistance of the workpiece. In this condition we would like the compliant mechanism to be stiff so that it can work effectively against the workpiece. Here, the input point A is assumed to be fixed, i.e. there is no additional applied load. An approach to handle multiple output ports is described in [4].

A multi-criteria optimization problem (Eq. (7.2)) is formulated to handle the trade-off between flexibility and stiffness using quantities called mutual potential energy (*MPE*) and strain energy (*SE*). *MPE* is used to specify the desired output deflection in terms of energy, which is a scalar quantity and therefore convenient to use in the objective function. Two equilibrium equations are solved for displacements u_A due to the actual load, and displacements v_B due to the virtual load, where K_1 is the stiffness matrix of the discretized finite element model. Here, f_A is the finite element load vector representation of applied load F_A, and similarly for the other load vectors. *MPE* can be thought of as the projection of virtual load vector f_B onto the actual displacements u_A. A third equilibrium equation is solved for displacements u_B, where K_2 is the stiffness matrix of the discretized finite element model, and the strain energy *SE* is calculated. Minimization of *SE* is equivalent to minimization of compliance in this case.

The objective function is formulated so as to maximize the ratio of *MPE* to *SE*, thereby simultaneously maximizing the output displacement in the desired direction and minimizing the *SE* (i.e. maximizing the stiffness) against the workpiece. Assumptions that are implicit in this approach include linear elastic material behavior and small deformations. Compliant mechanisms undergoing large deformations can also be considered, in which case large deformation finite element analysis must be employed. Constraints include an upper limit on the volume fraction F_v, and upper and lower limits on the design variables x_i. As an alternative approach to the second loading condition, a spring can be used at the output point to represent the stiffness of the workpiece [5, 6]. In any case, an output load or stiffness is needed; otherwise the optimizer has no motivation to connect material to the output point.

$$\max \left[\frac{MPE}{SE} \right] = \frac{v_B^T K_1 u_A}{u_B^T K_2 u_B}$$

$$s.t.$$

$$K_1 u_A = f_A$$

$$K_1 v_B = f_B \tag{7.2}$$

$$K_2 u_B = -f_B$$

$$F_v - \overline{F_v} \leq 0$$

$$\underline{x} \leq x_i \leq \overline{x}$$

To solve the topology optimization problem, the design must be parameterized. There are two main approaches to parameterization for topology optimization of compliant mechanisms, the ground structure approach and the continuum approach, which are described in the next two sections.

7.3 Ground Structure Approach

In the ground structure approach, a continuous design domain is approximated using a dense network of truss or beam elements. The design domain is discretized using nodes and the nodes are connected together by elements. The largest number of elements would be contained in a full ground structure, where every node is connected to every other node by an element. Whether a full ground structure or some subset of it is used, a large number of elements is usually needed to approximate a continuous structure.

An example of a compliant mechanism design problem solved using a ground structure of truss elements is pictured in Figure 7.5. The design domain is pictured in Figure 7.5a, which is a half-symmetry model of a compliant pliers problem. When the load **F** is applied, we would like the jaws of the pliers to move toward one another in the direction of Δ. The ground structure is pictured in Figure 7.5b. Here, the design variables are the cross-sectional areas of the truss elements. Topology optimization is accomplished by setting the lower limit on the design variables to a very small value, nearly zero. When the optimization converges, elements that have a value at or near this lower limit are considered to be void, and the remaining elements define the optimal topology. An intermediate solution is pictured in Figure 7.5c; here the optimizer is beginning to connect material to the support points, though many of the variables are near the lower limit. The optimal solution is pictured in Figure 7.5d. The elements that are at or near the upper limit are shown in black, the ones that are at or near the lower limit are not shown, and the rest are shown in relative shades of gray. This grayscale shading of elements is the standard method used to illustrate topology solutions. The deformation of the optimal solution is shown in Figure 7.5e.

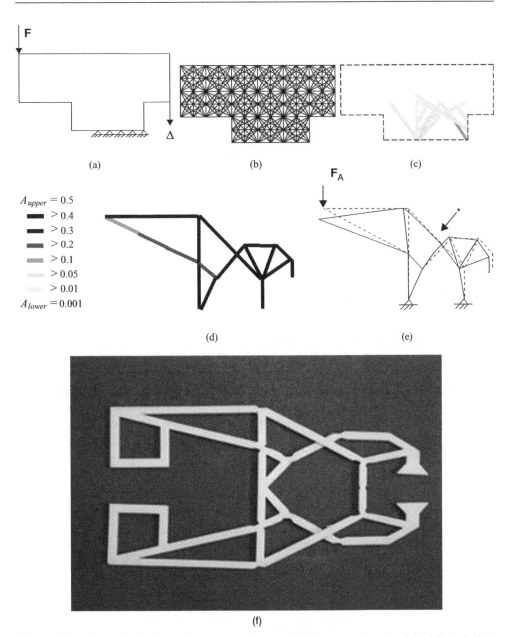

Figure 7.5 Compliant pliers design problem. (a) Half-symmetry model of the design domain with applied load *F* and desired output displacement Δ; (b) ground structure of truss elements; (c) intermediate solution; (d) optimal solution; (e) undeformed (dashed) and deformed (solid) solution; (f) prototype compliant pliers

It can be seen that the output point does displace in the vertical direction as desired, but there is also a component of the displacement in the horizontal direction. This example illustrates a limitation to the maximization of MPE/SE approach; although the output displacement is maximized in the desired direction, there is no direct control over the output displacement in any other direction. Strategies have been developed to address this issue and can be found in [7,8]. Another interesting artifact of the ground structure method is that there is the potential for overlapping elements, such as the elements pictured at point * in Figure 7.5e. In this example the overlapping elements are interpreted as a sliding joint. The presence of overlapping elements could make fabrication of a planar compliant mechanism more challenging. However, if all overlapping elements are removed from the initial ground structure, there are many fewer elements for the optimizer "to choose from". This trade-off would need to be handled by the designer, depending on the intended application. Finally, a compliant pliers prototype based on this topology solution is pictured in Figure 7.5f. This device was fabricated using fused deposition modeling rapid prototyping.

7.4 Continuum Approach

The continuum approach is another way to model the design domain in topology optimization problems. The approach typically uses a rectangular design domain that is discretized into quadrilateral finite elements. By using a fine mesh, the structural model more closely represents a continuum than a ground structure model. Here, two methods to parameterize the design domain are described, the solid isotropic material with penalization method and the homogenization method. The continuum approach as applied to compliant mechanism design is also described in [9].

7.4.1 SIMP Method

A widely used method to parameterize the design domain in topology optimization problems is called the solid isotropic material with penalization (SIMP) method. The SIMP method was developed by Bendsoe and Sigmund [7] and is summarized here. In this approach the relative densities, x^e, of each element are the design variables. A reference material of density ρ_0 is selected and the element densities ρ_e are calculated according to Eq. (7.3). If the relative element density x^e is equal to 1, then the element consists of solid material, whereas if the relative element density x^e is equal to the lower limit x^e_{min}, then the element is considered to be void. The lower limit x^e_{min} is set to a value very close to zero, but not equal to zero to avoid singularities in the stiffness matrix. An example of a design domain with solid (black), void (white), and intermediate (gray) elements is shown in Figure 7.6.

$$\rho_e = x^e \rho_0$$
$$0 < x^e_{min} \le x^e \le 1 \tag{7.3}$$

Elements with intermediate values of relative density (gray elements) can be interpreted by the designer as material or void. The SIMP method employs a strategy

Figure 7.6 In the SIMP method the design variables are the relative element densities. The example topology consists of solid elements (black), void elements (white), and intermediate elements (gray)

to avoid intermediate densities where a penalty factor p is used in the calculation of the element stiffness k^e, as in Eq. (7.4). Here, k^0 is the element stiffness matrix of an element consisting of the reference material. Using this technique, elements that have intermediate values of relative density are penalized and are uneconomical for the optimizer. A 99-line Matlab code has been developed by Prof. Ole Sigmund to solve the minimization of compliance problem using the SIMP method [10] and is available for download at [11]. More recently, the code has been reduced to 88 lines [12].

$$k^e = (x^e)^p \, k^0 \tag{7.4}$$

There have been a number of compliant mechanism topology design formulations that employ the SIMP method; one of the original formulations [6] is summarized here. The compliant mechanism topology design problem is pictured in Figure 7.7. In this approach springs are included at the input and output points A and B, respectively. The input spring could represent an actuator with stiffness k_{in}, and the output spring could represent a workpiece of stiffness k_{out}. The optimization problem is shown in Eq. (7.5), where the objective is to maximize the output displacement u_{out}, subject to

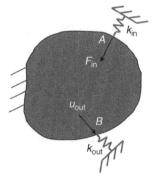

Figure 7.7 Compliant mechanism design problem. The output displacement u_{out} is maximized with input of stiffness k_{in}, and output spring of stiffness k_{out}

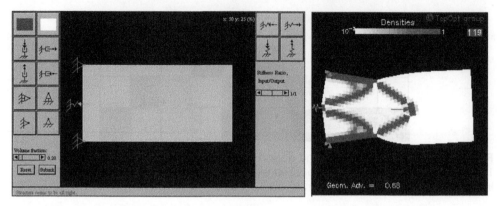

Figure 7.8 (Left) Compliant mechanism design example using the TopOpt Mechanism Design web tool (11). (Right) Displacement inverter solution with input force and output displacements in opposite directions

constraints on the total volume V, and upper and lower limits on the element densities ρ_e. Here, N is the total number of elements.

$$\max_{\rho} u_{out}$$
$$s.t.$$
$$\sum_{e=1}^{N} v_e \rho_e \leq V \tag{7.5}$$
$$0 < \rho_{\min} \leq \rho_e \leq 1, \quad e = 1, \ldots, N$$

A compliant mechanism topology design example solved using this formulation is shown in Figure 7.8. This problem was solved using the Mechanism Design Java applet on the TopOpt website [11]. The web tool allows the designer to place the input force and output deflection, as well as the support conditions, in a rectangular design domain. The user can also specify the volume fraction and any regions that should be nondesign material or void. For more information on the TopOpt web tool, the reader is referred to [13]. The example in Figure 7.8 illustrates the compliant displacement inverter problem, where the input force (shown in pink in the figure) and the output displacement (shown in blue) are in opposite directions. Notice that the optimal topology is very similar to the one shown in Figure 7.3a.

It is important to note that the solution pictured in Figure 7.8 (right) contains areas that act like hinges, where solid elements appear to be connected only at their corners. These "one-node connected hinges" appear because they allow mechanism-like behavior, which maximizes the output displacement. However, these one-node connected hinges can be practically undesirable because of the large stresses that may occur in a monolithic compliant mechanism with very thin hinge-like areas. Sigmund and coworkers [14–16] have recently developed a "robust formulation" for compliant mechanism design to ensure insensitivity to manufacturing variations and avoid

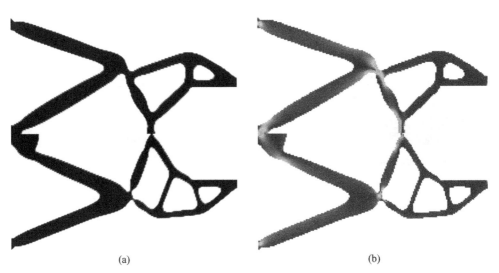

(a)   (b)

Figure 7.9 (a) (left). Compliant gripper topology solutions using robust (top) and standard (bottom) formulations. Figure 7.9b (right). Stress contours for compliant gripper topology solutions using robust (top) and standard (bottom) formulations. The robust formulation results in a solution with more distributed compliance than the standard formulation. Figures courtesy of Prof. Ole Sigmund

one-node connected hinges. Figure 7.9a shows the topology solutions for a gripper design problem for both the robust (top) and standard (bottom) formulations. The stress contours are pictured in Figure 7.9b. It can be seen that the locations of maximum stress occur in the thin hinge-like areas (shown in red) of the solution obtained using standard formulation (bottom), but that the stresses are more evenly distributed in the solution obtained using the robust formulation (top). Another desirable feature of the robust formulation is that it results in almost entirely black-and-white solutions, meaning that no post-processing is necessary to eliminate intermediate (gray) elements. Other compliant mechanism design issues that have been addressed while employing the SIMP method include control of the direction of the output control displacement, multiple outputs, and geometric nonlinearity [7].

7.4.2 Homogenization Method

Another widely used method to parameterize the topology design problem in a continuum approach is called the homogenization method. This method was originally developed by Bendsoe and Kikuchi [17] for minimum compliance design and was then applied to the compliant mechanism design problem [3], as well as many other structural design problems. In this approach the design domain is made up of a microstructure of unit cells consisting of material and void, as pictured in Figure 7.10. Rectangular holes are used with three variables per cell: μ, γ, and θ. If $\mu = \gamma = 0$, the cell is filled with material (solid), whereas if $\mu = \gamma = 1$, cell is completely void. Intermediate values of μ and γ define a porous structure. The orientation angle θ

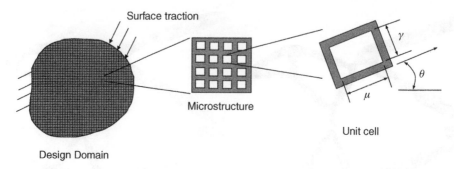

Figure 7.10 In the homogenization method, the design domain is modeled using a porous microstructure. There are three variables per unit cell that define the density and orientation of the cell

is usually defined to coincide with principle stresses. Optimal values of μ, γ, and θ are converted to a continuous density measure ρ defined as a function of the geometry of the holes in the microstructure and their orientation. The effective properties of the structure, or homogenized elasticity tensor (E_{ijkl}^{H}), is then calculated. E_{ijkl}^{H} are typically stored in a lookup table for certain hole sizes, and intermediate densities are typically handled using interpolation. Like the SIMP method, a penalty function can be used to avoid intermediate densities. Topology solutions using this method will be similar in appearance to those obtained using the SIMP method.

7.5 Discussion

It should be noted that the methods for parameterizing the design domain, i.e. ground structure and continuum approaches, are usually independent of the optimization problem formulations. The choice of parameterization method is made by the designer and depends primarily on computation time and software availability. The ground structure method may be preferred when computation time is important because a relatively small number of elements can be used and because it is relatively easy to develop the finite element analysis code. The homogenization method is more complicated and requires more computation time, and the homogenization formulas may not be available to the average designer trying to implement his/her own code. The SIMP method is convenient for many problems, in part due to the availability of the software on the Topopt website [11]. In any case, the designer should understand the underlying problem formulation and parameterization and any associated assumptions and limitations of these.

    A comparative study of the various optimization formulations for compliant mechanism design is described in [2]. Many different extensions of these formulations and other approaches have been developed by various researchers, and are too numerous to cite all of them here. Some examples include approaches accounting for large deformations [4, 8], material nonlinearity [18], dynamic applications [7, 19], self-contact [20–23], and manufacturing considerations [24]. The reader is referred to the sources in the list of references at the end of this chapter for additional reading on these and other approaches.

The main benefit of the topology optimization approach is that the designer need not start with a known mechanism. Topology optimization can generate novel solutions that the designer may not have come up with on his/her own, or it can be used to augment the designer's experience and intuition. In any case, the topology solution should be thought of as a rough outline of the optimal compliant mechanism. Whether the ground structure or continuum approach is used, the solution does require interpretation by the designer and post-processing. Often, further detailed design and finite element analysis is required to smooth out boundaries and avoid stress concentrations.

Limitations of topology design problems include the following:

- Optimal solutions are often mesh dependent.
- The problem is generally nonconvex: optimal solutions are generally not unique.
- The solution depends on value of material resource (volume) constraint, and on the starting point.
- Point flexures and lumped compliance are often present in topology solutions. The localized deformations and high stresses in these areas may not be ideal in practice.
- The topology solution is dependent on the designer's choice of output stiffness.
- The topology solution depends on the magnitude of the applied load(s) when geometric nonlinearity is taken into account.

7.6 Optimization Solution Algorithms

Various optimization solution algorithms are used in conjunction with compliant mechanism topology design. Gradient-based methods such as sequential linear programming and the method of moving asymptotes (MMA) are commonly employed, as are genetic algorithms (GA) and other heuristic methods. The choice of algorithm depends in large part on the nature and number of design variables. For the compliant mechanism design approaches described here, where there are a relatively large number of continuous variables, gradient-based methods are a good choice because they converge fairly quickly. A disadvantage of gradient-based methods is that when the objective and constraint functions are not explicit functions of the variables, the gradients are not easily calculated. Approximation methods such as the adjoint method or finite difference can be used, but this increases computation time significantly due to the large number of function evaluations required. Also, gradient-based methods tend to converge to a local optimum. This limitation can be addressed by doing numerous runs with many different starting points. On the other hand, heuristic methods such as genetic algorithms do not require calculation of gradients, and they converge to global optima, but these methods are best suited for problems with a relatively small number of design variables. GA can handle problems with discrete variables, or a combination of discrete-continuous variables. However, for problems with a large number of variables, the computation time becomes quite long. A comparative study of three approaches for topology design, cellular automaton method, the optimality criteria method, and the method of moving asymptotes is given in [25].

Acknowledgment

The author gratefully acknowledges Prof. Ole Sigmund of the Technical University of Denmark for providing the images shown in Figures 7.9a and b.

References

[1] Rozvany, G.I.N., ed. *Topology Optimization in Structural Mechanics.* 1997, CISM, Udine.

[2] Deepak, S.R., M. Dinesh, D.K. Sahu, and G.K. Ananthasuresh, A comparative study of the formulations and benchmark problems for the topology optimization of compliant mechanisms. *Journal of Mechanisms and Robotics*, 2009. **1**(1): p. 1–8.

[3] Frecker, M., G.K. Ananthasuresh, S. Nishiwaki, N. Kikuchi, and S. Kota, Topological synthesis of compliant mechanisms using multi-criteria optimization. *Journal of Mechanical Design, Transactions of the ASME*, 1997. **119**(2): p. 238–245.

[4] Saxena, A., Topology design of large displacement compliant mechanisms with multiple materials and multiple output ports. *Structural and Multidisciplinary Optimization*, 2005. **30**(6): p. 477–490.

[5] Saxena, A. and G. Ananthasuresh, On an optimal property of compliant topologies. *Structural and Multidisciplinary Optimization*, 2000. **19**(1): p. 36–49.

[6] Sigmund, O., On the design of compliant mechanisms using topology optimization. *Mechanics of Structures and Machines*, 1997. **25**(4): p. 495–526.

[7] Bendsoe, M.P. and O. Sigmund, *Topology Optimization: Theory, Methods and Applications.* 2003, Berlin Heidelberg: Springer-Verlag.

[8] Pederson, C.B.W., T. Buhl, and O. Sigmund, Topology synthesis of large displacement compliant mechanisms. *International Journal for Numerical Methods in Engineering*, 2001. **50**: p. 2683–2705.

[9] Ananthasuresh, G.K. and M. Frecker, Optimal Synthesis with Continuum Models, in *Compliant Mechanisms.* 2001, John Wiley and Sons.

[10] Sigmund, O., A 99 line topology optimization code written in Matlab. *Structural and Multidisciplinary Optimization*, 2001. **21**: p. 120–127.

[11] *TopOpt Research Group.* [cited April 2012]; Available from: http://www.topopt.dtu.dk.

[12] Andreassen, E., A. Clausen, M. Schevenels, B.S. Lazarov, and O. Sigmund, Efficient topology optimization in MATLAB using 88 lines of code. *Structural and Multidisciplinary Optimization*, 2011. **43**(1): p. 1–16.

[13] Tcherniak, D. and O. Sigmund, A web-based topology optimization program. *Structural and Multidisciplinary Optimization*, 2001. **22**(3): p. 179–187.

[14] Schevenels, M., B.S. Lazarov, and O. Sigmund, Robust topology optimization accounting for spatially varying manufacturing errors. *Computer Methods in Applied Mechanics and Engineering*, 2011. **200**: p. 3613–3627.

[15] Wang, F., B.S. Lazarov, and O. Sigmund, On projection methods, convergence and robust formulations in topology optimization. *Structural and Multidisciplinary Optimization*, 2011. **43**: p. 767–784.

[16] Lazarov, B.S., M. Schevenels, and O. Sigmund, Robust design of large-displacement compliant mechanisms. *Mechanical Sciences*, 2011. **2**: p. 175–182.

[17] Bendsoe, M.P. and N. Kikuchi, Generating optimal topologies in structural design using a homogenization method. *Computer Methods in Applied Mechanics and Engineering*, 1988. **71**: p. 197–224.

[18] Bruns, T.E. and D.A. Tortorelli, Topology optimization of non-linear elastic structures and compliant mechanisms. *Computer Methods in Applied Mechanics and Engineering*, 2001. **190**(26–27): p. 3443–3459.

[19] Maddisetty, H. and M. Frecker, Dynamic topology optimization of compliant mechanisms and piezoceramic actuators. *ASME Journal of Mechanical Design*, 2004. **126**(6): p. 975–983.

[20] Mankame, N.D. and G.K. Ananthasuresh, Topology optimization for synthesis of contact-aided compliant mechanisms using regularized contact modeling. *Computers & Structures*, 2004. **82**(15–16): p. 1267–1290.

[21] Mankame, N.D. and G.K. Ananthasuresh, Synthesis of contact-aided compliant mechanisms for non-smooth path generation. *International Journal for Numerical Methods in Engineering*, 2007. **69**(12): p. 2564–2605.

[22] Mehta, N., M. Frecker, and G. Lesieutre, Stress relief in contact-aided compliant cellular mechanisms. *ASME Journal of Mechanical Design*, 2009. **131**(9): p. 1–11.

[23] Reddy, B., S. Naik, and A. Saxena, Systematic synthesis of large displacement contact-aided monolithic compliant mechanisms. *ASME Journal of Mechanical Design*, 2012. **134**(1): p. DOI: 10.1115/1.4005326.

[24] Sigmund, O., Manufacturing tolerant topology optimization. *Acta Mechanica Sinica*, 2009. **25**(2): p. 227–239.

[25] Patel, N.M., D. Tillotson, J.E. Renaud, A. Tovar, and K. Izui, Comparative study of topology optimization techniques. *AIAA Journal*, 2008. **46**(8): p. 1963–1975.

8

Synthesis through Rigid-Body Replacement

Christopher A. Mattson

Brigham Young University, USA

This chapter describes one of the most useful and practical methods for compliant mechanism synthesis: rigid-body replacement. Procedures for and limitations of synthesis by rigid-body replacement are provided as well as a simple yet realistic synthesis example.

8.1 Definitions, Motivation, and Limitations

Before defining *rigid-body replacement* as a way to synthesize compliant mechanisms, let us first reiterate a few basic definitions already established in other parts of this handbook. Mechanisms are mechanical devices used to transfer or transform motion, force, and/or energy [1]. Linkages are some of the most common types of mechanisms. Rigid-body mechanisms have rigid links and movable joints. These movable joints are typically pin joints and slider joints. In contrast to rigid-body mechanisms, compliant mechanisms have at least one flexible link or one flexible joint. The pseudo-rigid-body model (PRBM) is a practical and convenient way to predict the kinematic performance of compliant mechanisms using traditional rigid-body mechanism analysis. The PRBM is presented in Chapter 5 of this handbook.

Synthesis is the process of designing a mechanism to accomplish a desired task [2]. Common mechanism tasks include path and function generation. One of the most practical and user-friendly ways to synthesize compliant mechanisms is through rigid-body replacement. In short, rigid-body replacement synthesis often begins with a rigid-body mechanism that is capable of accomplishing a desired task. The rigid-body mechanism is then transformed into a compliant mechanism by replacing rigid

Handbook of Compliant Mechanisms, First Edition. Edited by Larry L. Howell, Spencer P. Magleby and Brian M. Olsen.
© 2013 John Wiley & Sons, Ltd. Published 2013 by John Wiley & Sons, Ltd.

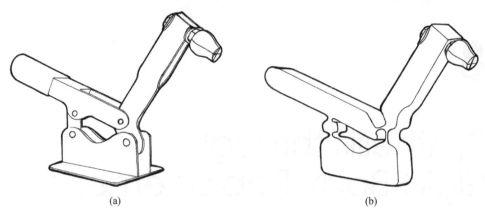

Figure 8.1   (a) Hold-down clamp (18 parts), (b) Compliant hold-down clamp concept (1 part)

links and movable joints with equivalent compliant members and joints. Importantly, traditional rigid-body mechanism analysis can be used to evaluate the performance of compliant mechanisms. The concept that ties rigid-body mechanism analysis to compliant mechanism analysis is the pseudo-rigid-body model. The links and joints of a rigid-body mechanism are used exactly as-is in the pseudo-rigid-body model for a compliant mechanism that can accomplish the same task. It's essential to recognize that many compliant mechanisms can be made from one pseudo-rigid-body model [1].

Compliant mechanism synthesis by rigid-body replacement is an ideal design approach for companies that have existing rigid-body mechanisms that they would like to transition into compliant mechanisms. For example, consider the hold-down clamp shown in Figure 8.1a. This rigid-body mechanism can, through relatively simply steps be made compliant, as shown in Figure 8.1b. As discussed later, compliant mechanism synthesis by rigid-body replacement can also be carried out in reverse without difficulty. In such cases we would start with a general compliant mechanism, identify its pseudo-rigid-body model, then use that PRBM as a rigid-body mechanism whose dimensions can be found such that the resulting mechanism achieves a desired task.

There are various reasons a designer may want to transition a rigid-body mechanism into a compliant mechanism. The compliant mechanism can often include fewer parts, involve less assembly and consequently can be less expensive to mass produce. Additionally, because compliant mechanisms often eliminate traditional pin joints, more precise motion can be achieved. This is of course due to the elimination of pin-joint backlash.

Before discussing the procedures for rigid-body replacement synthesis, we must consider the limitation of such synthesis. Not all rigid-body mechanisms will be feasibly converted to a compliant mechanism. Compliant mechanisms are often more constrained in terms of practical motion. For example, a rigid-body pin joint is free to continuously rotate, while a compliant flexure is not. Another limitation is that synthesis by rigid-body replacement will only lead us to compliant mechanisms that are characterized by the rigid body it is replacing. In other words, synthesis by rigid-body replacement does not identify new rigid-body mechanisms, it simply identifies many compliant mechanisms that can be used to replace the rigid-body mechanism.

8.2 Procedures for Rigid-Body Replacement

In this section, we discuss a procedure for synthesizing compliant mechanisms through rigid-body replacement. In the first section, we consider three synthesis scenarios and the procedure that can be used for each. In particular, we show how the library in this handbook facilitates rigid-body replacement. In the second section of the chapter, we consider how to choose the best configurations given loads, strains, and kinematics.

8.2.1 Starting with a Rigid-Body Mechanism

Assuming that we already have a rigid-body mechanism that has the desired motion, and that we want to convert this rigid-body mechanism into a compliant one, we can proceed through the following steps.

Step 1:   Identify the rigid-body model for the rigid-body mechanism under consideration.
            For example, consider the hold-down clamp again. A diagram of the rigid-body mechanism is shown in Figure 8.2. The rigid-body model that represents this mechanism shows that the output force (F_{OUT}) and rotation (γ) are a function of F_{IN}, θ, L_1, L_2, L_3, L_4 and the angles of such links. This relationship between inputs and outputs can be identified using traditional mechanism analysis [2]. Once identified, these relationships will become a fundamental part of the pseudo-rigid-body model.

Step 2:   Replace one or more of the rigid links and/or movable joints with equivalent compliant members

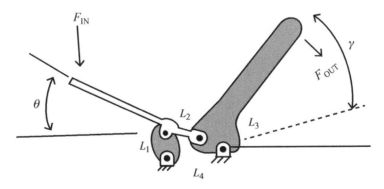

Figure 8.2   Rigid-body diagram

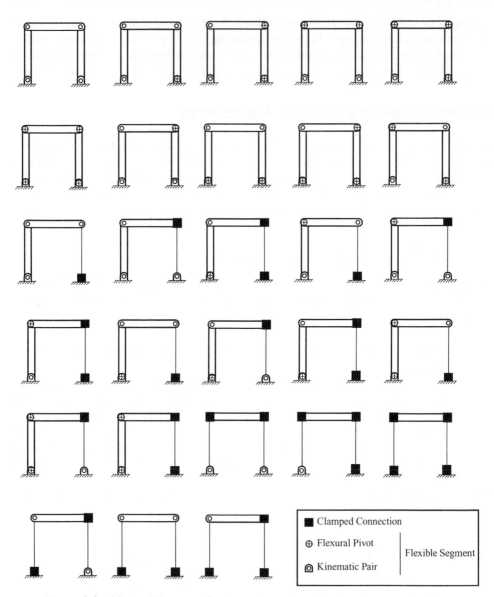

Figure 8.3   28 possible compliant configurations for four-bar mechanism

There are a variety of choices that can be made during this step of the procedure. For four-bar mechanisms, such as the clamp, there are 28 possible compliant mechanism configurations. These 28 configurations are determined by type synthesis [3, 4]. The 28 possible configurations are shown in Figure 8.3. Considering the hold-down clamp one possible rigid-body-replacement option is shown in Figure 8.4, where rigid links 1, 3 and 4, have been made one compliant piece.

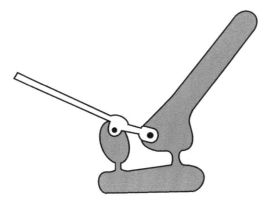

Figure 8.4   (a) One configuration for compliant clamp

While there are a variety of choices for this step, it is important to recognize that some configurations will be better than others. The choice of which configurations will be best while considering loads, strains and kinematics is more fully discussed in Section 8.2.4.

Step 3: Develop the pseudo-rigid-body model (see Chapter 5) for the selected configuration(s).

To develop the pseudo-rigid-body model for the compliant mechanism, we return to the rigid-body diagram and add the appropriate strain energy elements to the diagram. In the case of the hold-down clamp, two torsional springs are added as shown in Figure 8.5. These strain energy elements represent energy stored in the small-length flexural pivots. The tables found in Chapter 5 of this handbook show a variety of compliant mechanism segments, their pseudo-rigid-body representation, and the necessary mathematical equations such as the equations to determine the stiffness coefficients (K) from the geometry of the small-length flexural pivots.

Step 4: Select materials and size the compliant members to have desirable force-deflection relations and to withstand the resulting stress.

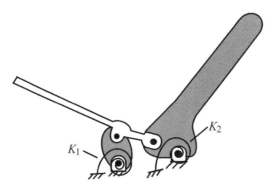

Figure 8.5   Pseudo-rigid-body diagram

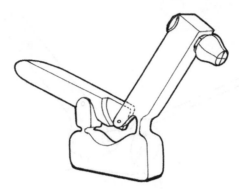

Figure 8.6 Resulting compliant hold-down clamp

The careful choice of materials, and of key geometry will lead to a compliant mechanism design that withstands the stress that stems from the force–deflection relationship. A resulting compliant version of the clamp is shown in Figure 8.6.

Steps 1–4 can create a compliant mechanism to match the motion and performance of an existing rigid-body mechanism. For the clamp shown in Figure 8.6, compliant joints replaced two of the movable joints. Rigid-body replacement synthesis has played an important role in this simple transformation.

8.2.2 Starting with a Desired Task

Under this scenario, we assume that we have a task that we want the compliant mechanism to perform, but we do not yet have a rigid-body mechanism that performs the desired task. To handle this scenario, we simply add Step 0 to the steps above.

Step 0:   Use one of many traditional mechanism synthesis methods to identify a rigid-body mechanism that is capable of performing the desired task.

Mechanism synthesis methods include type synthesis and dimensional synthesis. The goal of type synthesis is to identify which combinations of linkage topology and joint type are best suited for achieving the desired task. The exhaustive evaluation of combinations via type synthesis results in a set of combinations such as shown in Figure 8.3 for four-bar mechanisms. Dimensional synthesis is largely about determining the sizes of the links in order to achieve the desired task. The desired tasks are generally one of the following: multi-point motion generation, path generation, path generation with prescribed timing, and function generation [2].

Steps 1–4:   Follow the same Steps 1–4 listed in Section 8.2.1 above.

Steps 0–4 described directly above result in a compliant mechanism capable of achieving the desired task. A rigid-body mechanism was first identified to achieve the task, then that mechanism was simply transitioned to a compliant mechanism of equivalent performance.

8.2.3 Starting with a Compliant Mechanism Concept

There may be times that we have a basic compliant mechanism concept in mind and wish to size it properly so as to make it capable of achieving the desired task. Under such conditions, the following steps can be used.

Step 1: Develop the pseudo-rigid-body model for the compliant mechanism concept.

If a compliant mechanism concept is already selected, it must be determined if it (or some scaled version of it) will be able to accomplish the desired task. To do this we will develop the necessary models to predict the behavior of the compliant mechanism. The first model we need is the pseudo-rigid-body model, which can be derived by simply examining the elements of the compliant mechanism concept. From this, a pseudo-rigid-body diagram is made. For the clamp, Figure 8.1b shows a compliant mechanism concept, Figure 8.7 shows a diagram of the pseudo-rigid-body model. As can be seen in Figure 8.1b, small-length flexural pivots have replaced traditional pin joints. In Figure 8.7, these small-length flexural pivots have been modeled as torsional springs, with stiffness K_1–K_4.

Step 2: Extract the rigid-body model from the pseudo-rigid-body model.

The rigid-body diagram and model must now be extracted from the pseudo-rigid-body model. The rigid-body diagram for the clamp is shown in Figure 8.2.

Step 3: Use the rigid-body model in conjunction with traditional mechanism synthesis to identify the sizes for a rigid-body mechanism that is capable of performing the desired task.

To identify the sizes of the rigid-body links needed to accomplish the desired task, we use traditional mechanism synthesis methods; namely, dimensional synthesis [2]. The method discussed as Step 0 in Section 8.2.2 applies here.

Step 4: Select materials and size the compliant members to have desirable force-deflection relations and to withstand the resulting stress.

This is identical to Step 4 described in section 8.2.1.

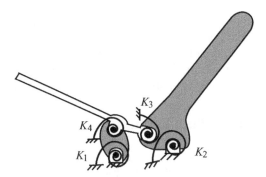

Figure 8.7   Pseudo-rigid-body diagram for compliant hold-down clamp

8.2.4 How Do We Choose the Best Configurations Considering Loads, Strains, and Kinematics?

Earlier, we mentioned that some compliant mechanism configurations are better than others given the conditions of the problem. The choice of which configurations will be best while considering loads, strains and kinematics is an essential part of rigid-body replacement synthesis. In this section, some basic design considerations are presented.

Consideration of Loads:
   The type of loading a mechanism's links and joints will experience will have a large influence on the configuration selection. Generally, links and joints will experience tensile, compressive, bending, and/or shear loading. As observed in this handbook, some joints are designed for only compressive loads (e.g., Library Element EM-48), while others for example are designed for both compressive and tensile loads (e.g., Library Element EM-11). Clearly any link undergoing compressive loads will need to be evaluated for safe buckling loads. Joints that remain as moveable pin joints need to undergo shear analysis.

Consideration of Strains:
   To understand how strain conditions affect configuration choice, consider the small length flexural pivot compared to the flexible link. The strain per unit length is noticeably larger in a small length flexural pivot then it is for a flexible link. If near or exceeding the yield limit in small length flexural pivots, a flexible link may prove to be better.

Consideration of Kinematics:
   The type of motion required for a mechanism's links and joints will also influence the configuration selection. For example, a flexible link can mimic the motion of rigid link only to about 60 degrees of rotation. If motion beyond this is needed, a pin-jointed rigid link may be best. Joining the consideration of kinematics, with the consideration of loads, we can find other viable options. For example, if the there is a link needing greater than 60 degrees of motion – and it will remain always in compression – a passive joint (Library Element EM-48) may be an excellent option.

Extending these considerations back to the hold-down clamp example, and examining the clamp's rigid body diagram (Figure 8.2) it can been seen that links 1 and 2 will be loaded in compression when the mechanism passes through its toggle point (link orientations at which links 1 and 2 are collinear). As such, configurations where these links are flexible were not considered. Flexible joints are selected, however, because they are more resistant to buckling. Nevertheless, care must be taken when sizing the small length flexural pivot so as to ensure that it will safely hold the loads.

8.3 Simple Bicycle Derailleur Example

In this section, a simple example is provided to illustrate the steps listed above. This derailleur example is motivated by the fact that manufactures of high-performance bicycles continually search for new design solutions that result in lighter bicycles – without compromising performance in other areas. Mass reductions on the order of grams can give manufacturers a significant competitive advantage. In a study by Mattson et al. [5], compliant mechanism technology was used to reduce the mass of a bicycle derailleur by 25 g (over 10% reduction in mass). In the study, one link in the four-bar derailleur was replaced with a compliant composite strip. Since the compliant strip stores potential energy when deflected, the tension spring that was part of the rigid-body derailleur was also removed. In the example presented in this chapter, we explore the basic compliant derailleur designed in Mattson et al. [5], and show how the steps of synthesis by rigid-body replacement were an integral part.

Design Problem Statement and Configuration Selection

Given, the Shimano Deore XT rear derailleur as a starting point, design a compliant bicycle derailleur that is lighter than the rigid-body Shimano derailleur of similar force–deflection characteristics.

Step 1:   Identify the rigid-body model for the mechanism under consideration.

The Shimano derailleur is shown in Figure 8.8a and is the mechanism under consideration. A rigid-body diagram of the four-bar derailleur mechanism is shown in Figure 8.8b. Notice that there is a tension spring that stores potential energy as the mechanism travels through its designed motion.

Four bar mechanisms are such that opposing links, which are of equal length, remain parallel throughout the motion of the mechanism. The traditional four-bar mechanism is shown as the top left figure of Figure 8.3. The parallel motion of the derailleur is used to position the chain on the rear sprockets of the bicycle.

The rigid-body model for the derailleur is found by analyzing the rigid-body diagram in Figure 8.9. Traditional kinematic analysis is used for this step and the designer

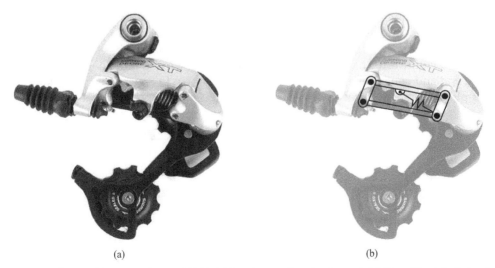

(a)   (b)

Figure 8.8   (a) Shimano Deore XT (rigid-body mechanism), and (b) Simplified sketch of the mechanism

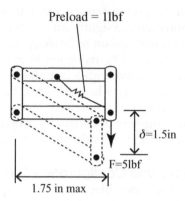

Figure 8.9 Rigid-body diagram for the bicycle derailleur with design criteria. This information is used to develop the rigid-body mathematical models

develops performance models for the items of interest. The performance measures typically include motion, force, and stress.

Step 2: Replace one or more of the rigid links and/or movable joints with equivalent compliant members.

For this step, we examine the rigid-body diagram (Figure 8.9) and we consider removing rigid and movable parts with compliant ones. As discussed earlier, type synthesis provides an exhaustive evaluation. Using type synthesis, 28 possible design configurations for the compliant four-bar derailleur were found and are shown in Figure 8.3. Considering the loading, strains, and kinematics, and using our best judgment regarding the feasibility of each configuration, we reduce the set of 28 designs down to two promising configurations. Specifically, we decide to maintain the load-bearing capabilities of the rigid-body derailleur, by retaining at least two rigid-body pin joints, and three rigid-body links in the compliant design. The two configurations that satisfy this requirement are shown in Figure 8.10. The basic configuration shown in Figure 8.10a has one compliant member (shown as a thin line), and has one fixed end and one pinned end. The configuration shown in Figure 8.10b also has one compliant member, however, both ends are fixed.

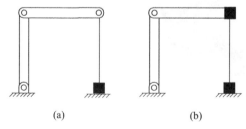

(a)   (b)

Figure 8.10 (a) Fixed-pinned compliant four-bar mechanism, and (b) fixed-fixed compliant four-bar mechanism

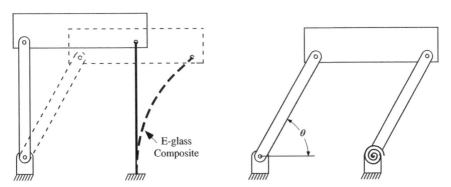

Figure 8.11 Fixed-fixed compliant four-bar mechanism configuration (left) and equivalent PRBM diagram (right)

Step 3: Develop the pseudo-rigid-body model (see Chapter 5) for the selected configuration(s).

For simplicity of presentation and for a practical reason discussed below, we will show how the pseudo-rigid body model is created for the fixed-fixed configuration shown in Figure 8.10b only.

We begin by creating a diagram of the pseudo-rigid-body; the right side of Figure 8.11 shows the pseudo-rigid-body model (PRBM) for the compliant derailleur concept.

After having the PRBM diagram, we develop the mathematical relationships for our performance measures. For the bicycle derailleur, the force, deflection, stress, and mass measures will be important. The basic modeling of a compliant fixed-fixed member is shown in Figure 8.12. For this configuration, the fixed parameters are $\delta = 1.5$ in, $\gamma = 0.8517$ [1], $K_\Theta = 2.6762$ [1], and the length of the rigid-body link (which needs to match the Shimano design) is given as $l_r = 1.75$ in, where l_r is γl_c.

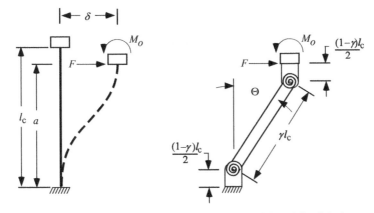

Figure 8.12 Pseudo-rigid body model for fixed-fixed flexible beam

The area moment of inertia is

$$I = \frac{bh^3}{12}$$

The length of the compliant member is

$$l_c = \frac{l_r}{\gamma}$$

The stiffness of the mechanism is evaluated as

$$K = \frac{2\gamma K_\Theta E I}{l_c}$$

The rigid-body angle, or the angle that the rigid link takes is

$$\Theta = \frac{\arcsin \delta}{\gamma l_c}$$

The vertical position of the end of the beam is given as

$$a = l_c(l - \gamma(1 - \cos \Theta))$$

The mass of the compliant member is

$$M = bhl_c\rho$$

where ρ is the density of the E-glass. The output force is

$$F = \frac{4K\Theta}{\gamma l_c \cos \Theta}$$

and the maximum bending stress is evaluated as

$$S = \frac{Fah}{4l^3}$$

The safety factor is

$$N = \frac{S_{max}}{S}$$

Step 4: Select materials and size the compliant members to have desirable force-deflection relations and to withstand the resulting stress.

An important part of compliant mechanism design is material selection. Materials with large S_y/E values are generally good candidates for compliant mechanisms. These materials are strong and flexible. One such material is E-glass composite; Young's modulus is 1 430 000 psi, yield strength in bending is 260 000 psi, and the mass density is 0.0931 lbm/in^3.

Considering the difficulty to create a pin joint with the E-glass composite, the fixed-fixed configuration shown in Figure 8.10b is selected as the basic configuration.

Given the mathematical relationships above, which can be found for any compliant mechanism using the developments in Chapter 5, the length, cross-sectional properties, and all other geometric parameters can be selected to achieved a desired outcome. Additional information about this example, including fatigue analysis and physical test results can be found in [5]. Numerical optimization method can also be used to evaluate concepts and identify good link geometry [6].

References

[1] Howell, L.L., *Compliant Mechanisms*. John Wiley & Sons, New York, NY, 2001.
[2] Erdman, A.G. and Sandor, G.N., *Mechanism Design: Analysis and Synthesis*, Volume 1, Prentice Hall, 1997.
[3] Murphy, M.D., Midha, A. and Howell, L.L. The topological synthesis of compliant mechanisms. *Mechanism and Machine Theory*, Vol. 31, no. 2, pp. 185–199, 1996.
[4] Derderian, J.M., Howell, L.L., Murphy, M.D., Lyon, S.M., and Pack, S.D. Compliant parallel-guiding mechanisms. Proceedings of the ASME Design Engineering Technical Conferences, DETC-1996-MECH-1208, 1996.
[5] Mattson, C.A., Howell, L.L. and Magleby, S.P., Development of commercially viable compliant mechanisms using the pseudo-rigid body model: case studies of parallel mechanisms. *Journal of Intelligent Material Systems and Structures*, Vol. 15, no. 3, pp. 195–202, 2004.
[6] Mattson, C.A., A New Paradigm for Concept Selection in Engineering Design Using Multiobjective Optimization, PhD Dissertation, Rensselaer Polytechnic Institute, 2003.

9

Synthesis through Use of Building Blocks

Charles Kim[a] and Girish Krishnan[b]

[a] Bucknell University, USA
[b] University of Michigan, USA

9.1 Introduction

Most engineered systems are designed using principles of modularity and often employ standardized components that provide specific functions. Standardized components such as motors, gear trains, and bearings may be treated as building blocks whose concatenation yields desired system behavior(s). There are myriad examples of engineered systems that can be decomposed into building blocks. Compliant mechanisms may be treated similarly by using mechanism building blocks to comprise a design and provide desired, overall functions [1–5].

In this chapter we present a general building-block approach for the design synthesis of compliant mechanisms. This building-block approach facilitates synthesis of original designs and may be implemented without an existing initial design. It is beneficial to use a building-block approach when a designer needs to generate a design "from scratch", although it is possible to use it when an initial design exists already. The designer is involved throughout the process, thereby increasing both insight and intuition. As a designer becomes more familiar with the capabilities of a set of building blocks, he/she better learns how to integrate the building blocks into an overall system for subsequent designs.

9.2 General Building-Block Synthesis Approach

The categorization of existing physical devices into sub-subsystems with specific functions is straightforward. Going the other way, however, is much more challenging.

Handbook of Compliant Mechanisms, First Edition. Edited by Larry L. Howell, Spencer P. Magleby and Brian M. Olsen.
© 2013 John Wiley & Sons, Ltd. Published 2013 by John Wiley & Sons, Ltd.

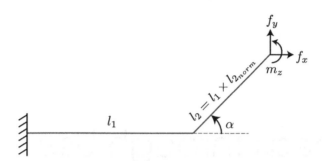

Figure 9.1   Compliant dyad building block

Given the overall function that a design must accomplish, it is often difficult to break down or *decompose* the overall function into *subproblems* that can be addressed by subsystems or building blocks.

A building-block synthesis approach presumes that a designer can decompose a given problem into more tractable functional subproblems. Decomposition is a nontrivial task that requires familiarity with overall system behavior and available building blocks that may meet the requirements of the subproblem. If this knowledge exists, however, a building-block approach can exploit an engineer's creative capacity. The design engineer can provide intuitive insight on how to intelligently decompose a problem and address the subproblems with multiple alternate solutions.

There are three major elements required for building-block mechanism synthesis: (i) a library of building blocks, (ii) models to characterize the primary functional behavior of a building block, and (iii) a means of functional decomposition[1]. The next three sections in this chapter offer an overview of these topics.

9.3 Fundamental Building Blocks

In this section, we present two fundamental building blocks, the compliant dyad and the compliant 4-bar. These two simple building blocks are capable of addressing a wide range of design problems, particularly when they are combined in series and parallel. In this section, we introduce these two fundamental building blocks and show how to incorporate them into designs in Section 9.5.

9.3.1 Compliant Dyad

The compliant dyad building block (Figure 9.1), CDB, is composed of two beams connected in series. The beams may assume different lengths (l_1 and l_2) and different orientations, as captured by α. The length of the second beam is normalized such that $l_{2_{norm}} = l_2/l_1$.

---

[1] For a general discussion of functional decomposition, see Otto and Wood [6]. Specific discussion of functional decomposition in reference to mechanism design can be found in [1–5].

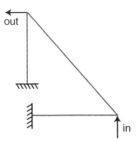

Figure 9.2   Compliant 4-bar building block

Most compliant mechanism designs consisting of beams are comprised of several CDBs. As we will see in subsequent sections, decomposing design tasks into CDB's aids in highlighting function and in making sense of mechanism geometry.

9.3.2 Compliant 4-Bar

The compliant 4-bar building block (C4B) is comprised of a CDB and an additional cantilevered beam (see Figure 9.2). The C4B has an input and an output, both of which are constrained to displace along specific directions. Such an arrangement of beams is particularly conducive to designing mechanisms that achieve displacement amplification.

9.4 Elastokinematic Representations to Model Functional Behavior

In this section, we present three methods to model the functional behavior of compliant building blocks. All of the models seek to extract fundamental elastokinematic behavior in a systematic manner that is valid independent of mechanism geometry.

Compliant mechanisms are fundamentally distinct from rigid-body mechanisms, because they obtain mobility through the deformation of their constituent elements. As a consequence, the kinematic behavior of a compliant mechanism is strongly coupled to the loads applied to it. This coupling is captured in a stiffness matrix, or its inverse, the compliance matrix.

$$F = [K]U$$
$$U = [C]F$$
$$[K] = [C]^{-1}$$

where

U = Generalized displacement vector
F = Generalized load vector
$[K]$ = Stiffness matrix
$[C]$ = Compliance matrix

Figure 9.3 Cantilever beam

The generalized displacement vector, U, captures both translations and rotations, while the generalized load vector, F, capture both forces and moments. Unfortunately, neither $[K]$ nor $[C]$ highlight fundamental functional behavior in and of themselves. Furthermore, traditional matrix decomposition methods (e.g. eigenvalue, Cholesky, etc.) do not yield meaningful information either[2]. In the next subsections, we present three geometric representations that capture fundamental mechanism behavior.

Consider the cantilevered beam shown in Figure 9.3. It is trivial to determine that the beam is most flexible in the transverse (y) direction. Describing this in a systematic fashion, however, is not straightforward. Under even a single force in the y-direction and no applied moment, the end of the beam translates in the y-direction and it also rotates.

We will consider the simple example of the cantilevered beam to demonstrate three mathematical characterizations that clarify the functional behavior of compliant building blocks. To that end, we will utilize the force–deflection relationship of a cantilevered beam captured in a 3×3 compliance matrix,

$$U = C_{3\times3}F = \begin{pmatrix} C_{11} & C_{12} & C_{13} \\ C_{21} & C_{22} & C_{23} \\ C_{31} & C_{32} & C_{33} \end{pmatrix} \begin{Bmatrix} f_x \\ f_y \\ m_z \end{Bmatrix} = \frac{1}{E} \begin{pmatrix} \frac{l}{A} & 0 & 0 \\ 0 & \frac{l^3}{3I} & \frac{l^2}{2I} \\ 0 & \frac{l^2}{2I} & \frac{l}{I} \end{pmatrix} \begin{Bmatrix} f_x \\ f_y \\ m_z \end{Bmatrix}$$

The compliance matrix is based on the deformation behavior of a linear elastic beam, which can be confirmed by considering the effect of the individual force and moment components on the deflection.

9.4.1 Compliance Ellipses and Instant Centers

A simple, yet insightful way to characterize a compliant building block is to consider the translation that results from an applied force in the absence of an applied moment. If a unit force is applied in various directions at a single point, we can determine the direction in which the point translates with the greatest magnitude. In the example of the cantilevered beam, we recognize this as the transverse y-direction.

More generally, one can apply a *unit force circle* to a single point. A unit force circle is transformed into an ellipse of displacement by the compliance matrix [1]. The ellipse is termed the compliance ellipse (Figure 9.4). The semi-major and semi-minor axes

---

[2] Mathematically, this is because the elements of the stiffness and compliance matrices do transform vectors of similar form. Physically, the difficulty is because forces and moments (or translations and rotations) cannot be directly compared to each other.

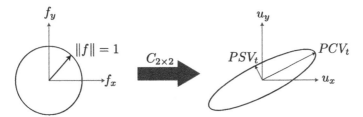

Figure 9.4   A unit force circle is mapped to an ellipse of displacement

of the ellipse identify the primary directions of linear compliance (PCV_t) and linear stiffness (PSV_t), respectively. The axes of the ellipse can be found as the eigenvectors of $C_{2\times2}$, the upper left 2×2 submatrix of C. The eigenvector that corresponds to the larger eigenvalue is PCV_t, while the other eigenvector is PSV_t.

9.4.1.1 Example: PCV$_t$ for a cantilevered beam

The compliance ellipse of a cantilevered beam is shown in Figure 9.5. The primary compliance vector, PCV_t, is in the transverse direction as expected. This implies that the beam is most flexible in the transverse direction. In all practicality, the compliance ellipse degenerates to a straight line. Its semi-minor axis is orders of magnitude smaller than its semi-major axis. This indicates that the beam is significantly stiffer in the axial direction.

9.4.1.2 Example: Instant Center of a Compliant 4-bar building block

The kinematics of a C4B can be characterized by recognizing that the PCV_t of each of the grounded beams constrains the input and output. If we assume that the floating beam remains relatively rigid, we can identify the *instant center* of the floating beam as shown in Figure 9.6. It follows that the relationship between the input and output translations can be captured as the geometric advantage $GA = \frac{u_{out}}{u_{in}} = \frac{l_{out}}{l_{in}}$.

9.4.2 Compliance Ellipsoids

The compliance ellipse captures primary linear compliance and stiffness but requires that loading be limited to pure linear force. Even under pure force, however, the resulting displacements are comprised of both translation and rotation. In contrast,

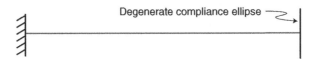

Figure 9.5   The compliance ellipse of a cantilevered beam tends to degenerate into a straight line because the transverse direction is much more compliant than the axial direction

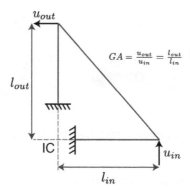

Figure 9.6 The instant center of the floating beam may be found as the intersection of the lines perpendicular to u_{in} and u_{out}

compliance ellipsoids capture the effect of both force and moment on the resulting displacement.

It is not possible to utilize standard matrix decomposition methods directly on C. This is because C must transform the generalized force vector $F = \{f_x\ f_y\ m_z\}$ to the generalized displacement vector $U = \{u_x\ u_y\ \theta\}$. The units of F are $[F\ F\ FL]$ while the units of U are $[L\ L\ 0]$. The compliance matrix must transform F to an entirely different type of vector.

To address this discrepancy, we introduce a normalizing length, l, to relate m_z to f_x and f_y and θ to u_x and u_y such that

$$U = \{u_x\ u_y\ \theta\} = \{u_x\ u_y\ {}^{u_\theta}/_l\}$$
$$F = \{f_x\ f_y\ m_z\} = \{f_x\ f_y\ lf_m\}$$

Typically, l assumes a nominal length of the scale of the building block. Rewriting the force-displacement relationships with respect to l, we obtain the following

$$U = \begin{Bmatrix} u_x \\ u_y \\ \frac{u_\theta}{l} \end{Bmatrix} = CF = C \begin{Bmatrix} f_x \\ f_y \\ lf_m \end{Bmatrix}$$

$$\begin{pmatrix} 1 & 0 & 0 \\ 0 & 1 & 0 \\ 0 & 0 & \frac{1}{l} \end{pmatrix} \begin{Bmatrix} u_x \\ u_y \\ u_\theta \end{Bmatrix} = C \begin{pmatrix} 1 & 0 & 0 \\ 0 & 1 & 0 \\ 0 & 0 & l \end{pmatrix} \begin{Bmatrix} f_x \\ f_y \\ f_m \end{Bmatrix}$$

$$\Rightarrow \tilde{U} = \begin{Bmatrix} u_x \\ u_y \\ u_\theta \end{Bmatrix} = \begin{pmatrix} 1 & 0 & 0 \\ 0 & 1 & 0 \\ 0 & 0 & l \end{pmatrix} C \begin{pmatrix} 1 & 0 & 0 \\ 0 & 1 & 0 \\ 0 & 0 & l \end{pmatrix} \begin{Bmatrix} f_x \\ f_y \\ f_m \end{Bmatrix} = \tilde{C}\tilde{F}$$

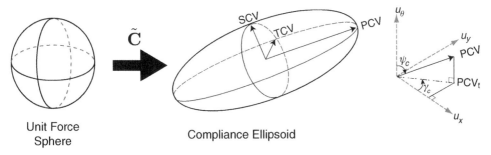

Unit Force
Sphere   Compliance Ellipsoid

Figure 9.7 The normalized compliance matrix transforms a unit force sphere to the compliance ellipsoid. The direction of the primary compliance vector can be described by two angles (ψ, γ)

The normalized compliance matrix, $\tilde{C}$, transforms $[\tilde{F}] = [F,F,F]$ to $[\tilde{U}] = [L,L,L]$. This transformation may be decomposed using traditional matrix methods because the units are consistent.

$\tilde{C}$ transforms a unit force *sphere* to a *compliance ellipsoid* (Figure 9.7). The semi-axes of the ellipsoid are found as the eigenvalues and eigenvectors of $\tilde{C}$. We identify the primary compliance vector (PCV), secondary compliance vector (SCV), and tertiary compliance vector (TCV) as seen in Figure 9.7. The PCV is the primary displacement direction, while the TCV is the primary constraint direction.

The three most salient characteristics that compliance ellipsoids identify are:

- γ: The angle between u_x and the projection of PCV onto the $u_x - u_y$-axis.
- ψ: The angle between the u_θ-axis and the PCV. ψ measures the coupling between rotational translation components in the PCV. As $\psi \to 0°$, the PCV becomes purely rotation. Conversely, as $\psi \to 90°$ the PCV becomes purely translational.
- n_2: The ratio $|\text{SCV}| / |\text{PCV}|$. This ratio indicates how much stiffer SCV is compared to PCV. If $n_2 \to 0$, the PCV is the dominant degree of freedom, while all other directions are constrained.

9.4.2.1 Example: Compliance Ellipsoid for a compliant dyad

The CDB building block can assume a wide variety of ellipsoid characteristics by varying only 2 parameters ($l_{2\text{norm}}$ and α). Figure 9.8 shows values for γ, ψ, and n_2 ($l_1 = 60$ mm, $l = 10$ mm). In the plots, the polar coordinates correspond to $(r, \theta) = (l_{2\text{norm}}, \alpha)$.

As $|\alpha|$ increases, the CDB becomes less "beam-like". The ratio of translation to rotation decreases in the PCV, thus resulting in a decrease in ψ. In general, as $|\alpha|$ increases, n_2 also increases, indicating that the building block begins to assume an additional degree of freedom.

Three building-block geometries are shown with their corresponding ellipsoid characteristics in Figure 9.9. The straight beam behaves in a predictable manner. The translational part of the PCV is in the transverse direction and the PCV consists of a small rotational component ($\psi = 82.8°$). Furthermore, the SCV is 200 times stiffer than

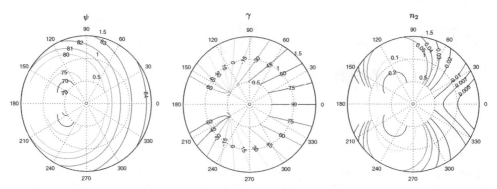

Figure 9.8 Plots of ψ, γ, and n_2 for a CDB ($l_1 = 60$ mm, $l = 10$ mm). Polar coordinates in the plots correspond to $(r,\theta) = (l_{2norm}, \alpha)$

the PCV ($n_2 = 0.005$). As α increases, this behavior changes drastically. The primary direction of compliance becomes less dominant as n_2 increases, while the ratio of rotation and translation in the PCV increases as ψ decreases. These trends may be observed in the plots in Figure 9.8 and in the specific geometries shown in Figure 9.9.

9.4.3 Eigentwist and Eigenwrench Characterization

In the previous section a representation of planar compliance at a single point was captured by three-dimensional compliance ellipsoids. The use of an arbitrary normalizing length, however, compromises the mathematical robustness of this representation. Alternatively, the *eigentwist and eigenwrench* characterization decouples translational and rotational terms from the compliance matrix without introducing

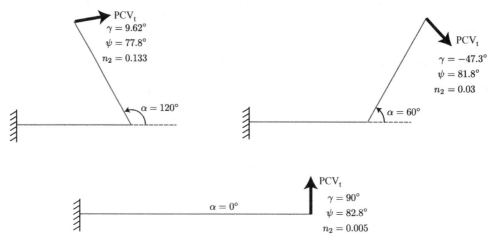

Figure 9.9 Compliance ellipsoid characteristics for three distinct CDB geometries ($l_1 = 60$ mm, $l = 10$ mm)

a normalizing length. The resulting characterization is intrinsic to the geometry and leads to an insightful functional characterization. This enables the building block method to be accomplished graphically using intuitive geometrical entities.

9.4.3.1 Decoupling translations and rotations

The eigentwist and eigenwrench decomposition is defined by two generalized eigenvalue problems

$$CF = a_f \tilde{\eta} F \quad KU = k_g \tilde{\xi} U$$

where C and K are the compliance and stiffness matrices, respectively, and U and F are the generalized displacement and force vectors. $\tilde{\eta}$ and $\tilde{\xi}$ selectively normalize the translations and the rotational terms of the compliance and stiffness matrices, respectively, and take the form

$$\tilde{\eta} = \begin{bmatrix} I & 0 \\ 0 & 0 \end{bmatrix} \quad \tilde{\xi} = \begin{bmatrix} 0 & 0 \\ 0 & 1 \end{bmatrix}$$

where I is the identity matrix. The eigenvalues a_f contain translational compliance parameters alone while k_g contains values for rotational compliance.

9.4.3.2 Description of the Eigentwist and Eigenwrench parameters (3)

While the decomposition presented above is successful in decoupling translational and rotational terms in the compliance matrix, the associated parameters must be mapped to the mechanism geometry to enable insightful characterization. A more detailed discussion on how to obtain these parameters from the eigentwist and eigenwrench characterization is presented in Krishnan et al. [3]. These parameters are introduced below.

(a) *Center of Elasticity:* One of the key features of planar geometries is the existence of a unique point known as the center of elasticity, where translations and rotations are decoupled. At the center of elasticity (CoE) any force applied leads to pure translation alone, assuming a rigid connection with the input. Furthermore, any moment at this point leads to pure rotation. Thus, the distance (r_E) and orientation ($\beta + \delta$) of the CoE from the input as seen in Figure 9.10a are the preliminary geometrical parameters used to characterize compliance. The ratio of the distance r_E normalized by the length of a dyad building block is plotted in Figure 9.10d for varying dyad length ratios and angles. It must be noted that when $r_E = 0$, the CoE coincides with the input, and translational and rotational compliances are inherently decoupled.

(b) *Translational Compliance:* At the CoE, there exist two mutually perpendicular directions in which any force applied leads to a purely coincident translation. One direction is compliant, while the other is stiff. The compliant direction is captured by an angle δ with respect to the horizontal and its compliance is given by a_{f1}.

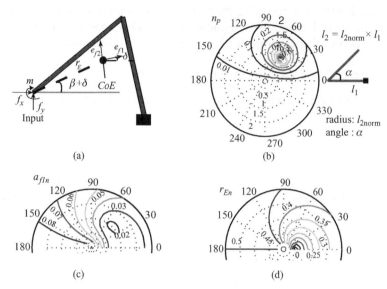

Figure 9.10 Eigentwist and eigenwrench parameters for a compliant dyad and its parametric variation with respect to the dyad geometry

Similarly, the compliance in the stiff direction is given by a_{f2}. A plot of $n_p = a_{f2}/a_{f1}$ is shown in Figure 9.10b as a function of the dyad angle (α) and the ratio of the length of the dyads $l_{2\text{norm}} = l_2/l_1$. a_{f1} normalized by l^3/EI is plotted in Figure 9.10c for various dyad geometries.

(c) *Eigenrotational Stiffness* (k_g): This gives the reaction moment produced by a pure unit rotation at the CoE. The eigenrotation stiffness is given as the ratio of the flexural rigidity (EI) to the overall length ($l_1 + l_2$), where I is the second area moment and E the Young's modulus of the material.

The compliance can thus be represented by the eigentwist and eigenwrench characterization using six geometrically relevant parameters – r_E, β, δ, a_{f1}, a_{f2}, and k_g. However, these parameters by themselves may not aid in insightful decomposition of a problem specification, which is a key step in building-block synthesis. To enable this, we present a graphical representation of the parameters. The details of obtaining the graphical representation from the compliance and stiffness matrices are explained in Krishnan et al. [3].

9.4.3.3 Graphical Representation of Compliance

The compliance matrix at the input consists of translational terms in its upper left 2×2 entries. This is represented by an ellipse with semi-major axis a_{f1} inclined at an angle δ from the horizontal as shown in Figure 9.11a. This ellipse characterizes compliance at the center of elasticity. To relate it back to the input, an additional degenerate ellipse of magnitude r_E^2/k_g and inclined at $\beta + \delta$ to the horizontal must be added. The off-diagonal terms (C_{13} and C_{23}) can be represented as a *coupling vector*

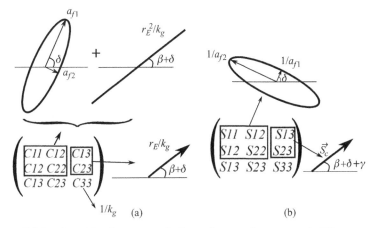

Figure 9.11 Graphical representation of compliance and stiffness matrix

that denotes the coupling between translational and rotational terms. The term C_{33} is the inverse of the rotational stiffness $(1/k_g)$. Similarly, stiffness at the input can be represented as the inverse of the compliance ellipse and a stiffness coupling vector (see Figure 9.11b) whose magnitude and orientation is given as

$$s_c = \frac{1}{a_{f1}} \left(\frac{1}{n_p^2} \cos^2 \beta + \sin^2 \beta \right)$$

$$\gamma = \tan^{-1} \left(\frac{1 - n_p}{(1 - n_p) \cos 2\beta + 1 + n_p} \right)$$

where $n_p = \dfrac{a_{f2}}{a_{f1}}$.

9.4.3.4 Graphical Depiction of Series and Parallel Combination

In the previous section, the eigen-twist and eigen-wrench parameters were used to propose a graphical representation of compliance at a single port. However, the representation itself does not enable insightful decomposition of a problem specification. The only two ways such a decomposition can occur is by dividing into subproblems that are connected in series or parallel. We thus present series and parallel combination in terms of the graphical representation of compliance.

Series Combination

Consider two building blocks, BB1 and BB2, of Figure 9.12a1 in series. The mechanics of series combination dictates that the *coupling vectors of the two building blocks add*. However, in series combination of BB2 on BB1, the input shifts from the latter to the former. This is accounted for by evaluating the coupling vector of BB1 from the input point I_{p2} as $\bar{c}_n = r_1/k_{g1}$. This is shown in Figure 9.12a2.

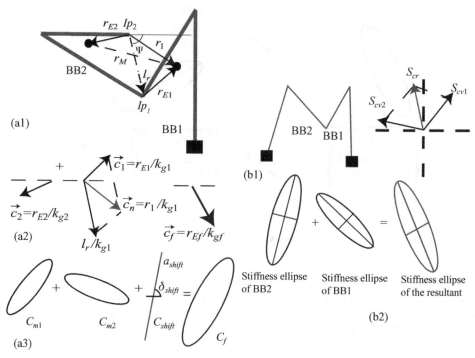

Figure 9.12 Graphical depiction of series and parallel combination

The compliance ellipses of the individual building blocks add in series combination. Since the compliance ellipse of each building block is evaluated at the CoE, there is an additional degenerate ellipse a_{shift} that is added that is indicative of the distance between CoEs of the individual building blocks.

$$a_{shift} = \frac{r_m^2}{k_{g1} + k_{g2}}$$

This degenerate ellipse is oriented perpendicular to the line joining the CoEs of the building blocks as seen Figure 9.12a3.

Parallel Combination

When building blocks are combined in parallel, stiffness ellipses and the stiffness coupling vectors add as seen in Figure 9.12b. Because the building blocks combine at the same point, there is no need for a shift ellipse as in the case of series combinations.

9.5 Decomposition Methods and Design Examples

The characterizations presented in Section 9.4 highlight the functional behavior of compliant mechanisms and the building blocks that comprise them. The real utility of these characteristics, however, is that they enable *functional decomposition*. Functional decomposition involves dividing desired mechanism behavior into several tractable subproblems that can more easily be addressed by available building blocks. In this section, we describe several methods to decompose desired mechanism behavior to enable systematic building-block design synthesis. The reader is encouraged to explore these decomposition methods in references [1–4].

9.5.1 Single-Point Mechanisms (3)

The eigentwist and eigenwrench parameters enable a geometric representation of compliance at a single point using the compliance ellipse and coupling vector. There is a direct mapping between the building-block geometry and its compliance parameters. Furthermore, the depiction of series and parallel combinations as addition of vectors and ellipses enables insightful problem decomposition. The prerequisite, however, is to represent the problem specification in terms of these geometric quantities. Consider the following example that illustrates the usefulness of the representation and decomposition strategies.

9.5.1.1 Design Example: Vision-based force sensor

Automation of manipulation tasks in the micro- and mesoscales require high-resolution sensors that can operate in various environments that are considered unfavorable for conventional piezoelectric and piezoresistive counterparts. A cost-effective solution involves a compliant mechanism, which is integrated in series with the actuator. As the actuator applies force on the object to be manipulated, the compliant mechanism deforms. The magnitude and direction of the deformation can be mapped to the amount of force applied. Such a planar sensor must preferably have equal stiffness to loads applied in any direction within the plane implying a circular compliance ellipse. We will aim to achieve a specific stiffness value of 0.15 mm/N. The mechanism will be made out of steel spring sheet of thickness 10 mm. Furthermore, for the ease of visual read-out, a force applied must produce a pure translation while a moment applied must produce a pure rotation. This requirement corresponds to having a coupling vector with a *zero* magnitude. First, we shall aim to meet the problem specifications by series combination of two building blocks.

Procedure for series combination
1. *Graphical Problem Specification*. The problem specification described above can be represented as a *circular* compliance ellipse and *zero-length* coupling vector as shown Figure 9.13a.
2. *Estimate Eigenrotational Stiffness*. The eigen-rotational stiffness signifies the response of the mechanism to a moment load. If the problem does not specify the eigen-rotational stiffness, strategies to select the appropriate value based on stress considerations and footprint have been proposed in Krishnan et al. [3]. Based on such a consideration, we fix the value to be 3.75×10^4 N mm. By making another simplifying assumption that the individual building blocks have equal eigen rotational stiffness, the value of $k_{g1} = k_{g2} = 7.5 \times 10^4$ N mm.

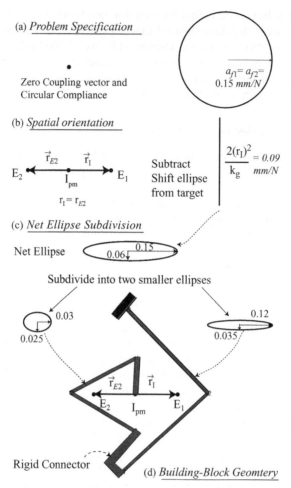

(a) *Problem Specification*

Zero Coupling vector and
Circular Compliance

$a_{f1} = a_{f2} = 0.15\ mm/N$

(b) *Spatial orientation*

$\vec{r}_{E2}$ $\vec{r}_{I}$

$E_2 \longleftarrow \bullet \longrightarrow E_1$

I_{pm}

$r_{I} = r_{E2}$

Subtract
Shift ellipse
from target

$\dfrac{2(r_I)^2}{k_g} = 0.09\ mm/N$

(c) *Net Ellipse Subdivision*

Net Ellipse   0.06↓ 0.15

Subdivide into two smaller ellipses

0.03

0.025

0.12

0.035

$\vec{r}_{E2}$ $\vec{r}_{I}$

E_2 I_{pm} E_1

Rigid Connector

(d) *Building-Block Geomtery*

Figure 9.13 Steps involved in designing for compliance with series combination of building blocks

3. *Spatially Orient the CoE*: For the present problem, since the coupling vector magnitude must be *zero*, the coupling vectors of the individual building blocks must be orientated such that they are equal and opposite. Since the coupling vector is oriented along the direction of the line joining the input and the CoE, the CoEs of the individual building blocks must be equidistant from the input as shown in Figure 9.13b.

4. *Evaluate the Net Ellipse*: It was seen earlier that series combination involved addition of the compliance ellipse of individual building blocks in addition to a shift ellipse proportional to the square of the distance between their CoEs. Having fixed the location of the CoEs in the previous step, the shift ellipse is evaluated as shown in Figure 9.13b. Care must be taken to ensure that the magnitude of the shift ellipse is less than the required compliance magnitudes. In this case choosing

$r_{E2} = r_I = 60$ mm ensures a shift ellipse value of 0.09 mm/N, well under the a_{f1} requirements of the problem.

5. *Net Ellipse Subdivision*: Having subtracted the shift ellipse from the required compliance ellipse we are left with the net ellipse shown in Figure 9.13c. The net ellipse can be further subdivided into two smaller ellipses that correspond to the individual building blocks. For simplicity the orientations of both the building blocks have been chosen in the same direction of the net ellipse. The dimensions of the dyad that corresponds to these building blocks are chosen as shown in the next step. In practice, these two steps must be iterative in nature in order to determine the optimal subdivision that yields a mechanism within the footprint requirements, prevent overlap of elements between the two building blocks, and minimize the number and size of rigid elements required to connect them.

6. *Building-Block Design and Combination*: Individual building-block geometries (l_{2norm} and α) are determined based on the ratio n_p between the secondary and primary eigencompliance magnitudes from Figure 9.10b. For the two ellipses the dyads were chosen such that their $l_{2norm} = 1$ and α are 57° and 87° respectively. By finding the normalized eigencompliance values for the two geometries from Figure 9.10c, the cross sections and the overall dyad length can be designed to meet the actual values of its eigen-compliance and eigen-rotational stiffness. The resulting two dyads shown in Figure 9.13c are oriented so that their CoEs are placed at the locations decided in step 3. If the ends of the building blocks do not meet, a rigid member is used to connect them. Furthermore, a rigid member is used to connect the input point from the input of the first dyad. The resulting geometry has equal bidirectional stiffness and completely decoupled translational and rotational compliance. Iteration between steps 5 and 6 is recommended to find the optimum building-block geometries without the need for rigid members.

Though one can obtain any compliance specifications by series combination of building blocks alone, the CoE is always bound within the smallest rectangle that encloses the mechanism [3]. This restricts the available space around the input. We propose a blend of series and parallel combination of building blocks to obtain greater freedom in placing the CoE.

Procedure for Parallel Combination

During parallel combination, the stiffness coupling vectors and ellipses of individual building blocks add without the need for shift ellipses. For this example, requiring circular stiffness ellipse and zero coupling vector can be achieved by combining symmetric halves of a submechanism. If the symmetric halves are oriented such that their coupling vectors are equal and opposite, they cancel out. Furthermore, if the symmetric halves have equal or circular compliance, the orientation of their stiffness coupling vectors becomes equal to the compliance coupling vector with $\gamma = 0$. Symmetric halves with equal compliance can be obtained by following the steps for series combination with arbitrary *CoE* orientation of step 3. One such submechanism is shown in Figure 9.14a. The stiffness coupling vectors can be oriented such that they make an angle with the horizontal by rotating the symmetric halves

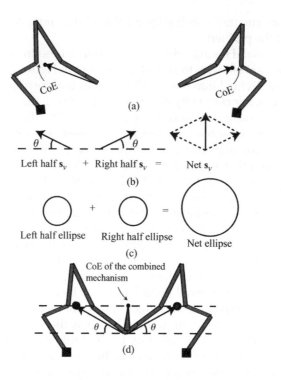

Figure 9.14 Demonstrating parallel combination of symmetric halves

as shown in Figure 9.14. This enables the use of a rigid probe that connects the input and the CoE to enable manipulation. The net stiffness ellipse of the mechanism will be a sum of the individual building-block ellipses.

9.5.1.2 Design Example: Rectilinear Constraint

A compliant constraint having two degrees of freedom is conceptualized as a straight beam in constraint-based design method. However, a beam has significant coupling between the two degrees of freedom, i.e. transverse displacement and out-of-plane rotation. This is known as cross-axis error. A parallelogram configuration reduces this error, but theoretically does not eliminate it. This is seen in the Figure 9.15a, which shows the center of elasticity (center of stiffness) to be at a distance away from the input. This distance indicates the amount of rotation a unit force produces. We present a solution that eliminates this error completely, for small displacements, using the design methodology based on eigen-twist and eigen-wrench parameters.

A requirement for having no rotation for a linear force would result in a zero resultant coupling vector, making the center of elasticity (center of stiffness) coincide with the input. Using two building blocks, the coupling vector of both the building blocks should cancel. We see that two beams connected by a rigid body is capable of

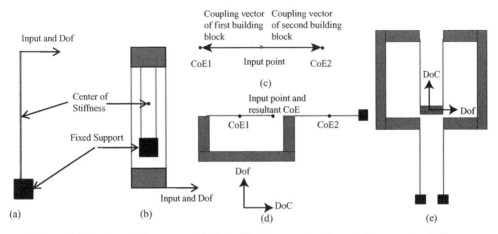

Figure 9.15 A rectilinear constraint with decoupled translations and rotations

having its CoE at the input. This is shown in Figures 9.15c and d. One can make the design symmetric to make it robust as shown in Figure 9.15e. Such a design could be used for MEMS suspensions and as building blocks for a compliant X-Y stage.

9.5.2 Multi-Port Mechanisms using Compliance Ellipsoids (4)

It is possible to design constraints where load is applied at a single point using either eigentwist or compliance ellipsoid characterizations. Designing mechanisms that consist of multiple ports, however, requires significantly more effort to decompose a problem. General specifications are shown in Figure 9.16a for a single-input–single-output mechanism with required input and output motion. This problem can be decomposed into three mechanism subproblems – (i) an input constraint, (ii) an intermediate submechanism, and (iii) an output constraint (Figure 9.16b). Both the input and output constraint subproblems can be addressed directly using building blocks characterized by compliance ellipsoids or eigentwist parameters. The intermediate submechanism, however, requires additional information.

The primary function of the intermediate transmission submechanism is to transmit load between the input and output ports. One way to do this is to align the primary

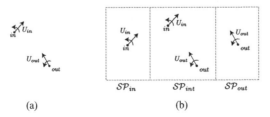

Figure 9.16 (a) Two-port motion specifications. (b) Decomposition into input constraint, intermediate submechanism, and (b) output constraint subproblems

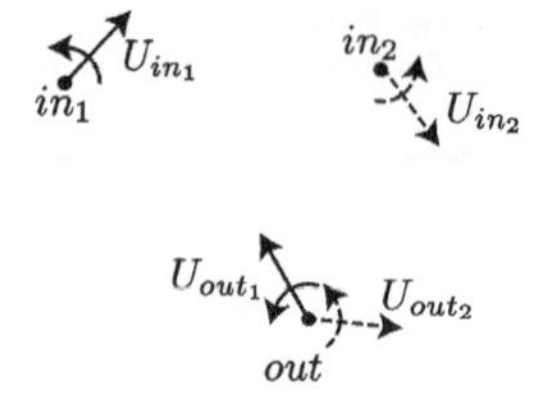

(a) DISO Motion Specifications

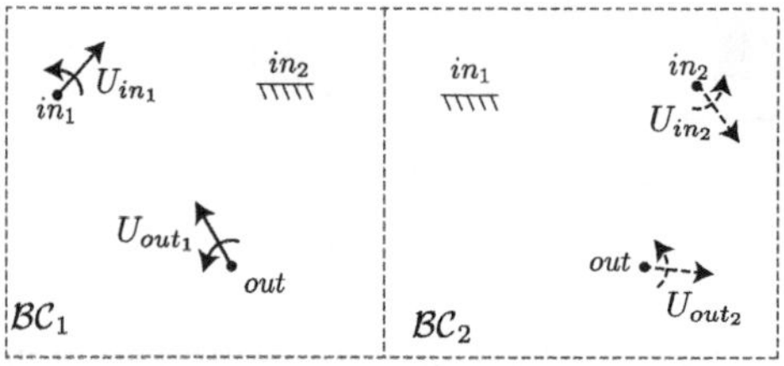

(b) Boundary Conditions for SISO Sub-problems

Figure 9.17   DISO motion specification and loading conditions

stiffness direction with the direction of the output load. Note that the primary stiffness direction is parallel with the translational part of TCV of a compliance ellipsoid. We define the *compliant deviation angle*, δ, as the angle between the TCV and the desired output displacement, U_{out}. The intermediate submechanism may be selected by ensuring that δ is minimized.

Decomposing a problem into input and output constraints and intermediate transmission submechanisms can be utilized for problems more complex than the SISO problem shown in Figure 9.16. One more complex problem is for the design of dual-input–single-output (DISO) mechanisms. With this type of mechanism, it is possible to obtain complex motions at the output of a compliant mechanism because it is driven by two distinct inputs. A DISO motion specification can be expressed as two loading conditions in which the individual actuators are alternately active or fixed (Figure 9.17). In practice, both of the actuators may be active to provide a *locus* of output displacements.

The decomposition of the DISO problem is shown in Figure 9.18. The output constraint subproblems (SP$_{out1}$, SP$_{out2}$) overlap with the interior mechanism subproblems (SP$_{int2}$, SP$_{int1}$). These building blocks must serve as both constraints and transmission submechanisms. It is important that the solutions to these subproblems are consistent. The constraint subproblems (SP$_{out1}$ and SP$_{out2}$) dictate that the PCV of the output constraints be parallel to the desired output directions. That is:

$$BB_1: PCV\|U_{out1}$$
$$BB_2: PCV\|U_{out2}$$

The building blocks, BB$_1$ and BB$_2$, must also act to transmit displacement for SP$_{int2}$ and SP$_{int1}$. To effectively transmit displacement to the output, the direction of primary

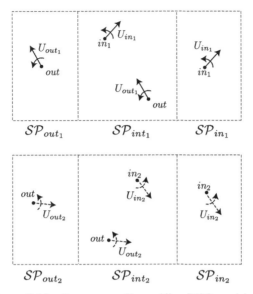

Figure 9.18   Decomposition of the DISO problem

stiffness (PSV or TCV) of the building blocks must be parallel to the *alternate* desired output direction. That is:

$$BB_1: TCV_{int2}||U_{out2}$$
$$BB_2: TCV_{int1}||U_{out1}$$

Together, the two building blocks must satisfy the following:

$$BB_1: PCV||U_{out1} \text{ and } TCV_{int2}||U_{out2}$$
$$BB_2: PCV||U_{out2} \text{ and } TCV_{int1}||U_{out1}$$

This decomposition will be demonstrated in the following design example.

9.5.2.1 Design Example: Dual-Input Gripper

In this problem, we seek to design a differential gripping mechanism with two possible output motions driven by two distinct input points. The desired motions and available design space are shown in Figure 9.19. The input motions are both in the $- x$-direction. The desired output translations are

$$\vec{u}_{out_1} = 10e^{-i90°} \text{ mm}$$
$$\vec{u}_{out_2} = 10e^{-i160°} \text{ mm}$$

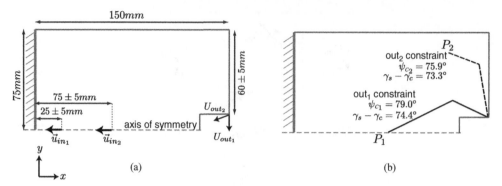

Figure 9.19 (a) Motion specifications for the differential gripper. (b) Output constraints selected to also provide load bearing for the corresponding interior transmission subproblem

The rotation of the output displacement is not specified but is limited to a maximum of 15° rotation per 10 mm translation for each loading case. The motion specifications translate to desired ellipsoid characteristics such that:

Output constraints and intermediate submechanisms
The DISO problem is decomposed into two subproblems (SP_{out1} and SP_{out2}) for the output constraints for each loading case. The output constraints must also serve as the intermediate transmission submechanisms (SP_{int1} and SP_{int2}). It is imperative that the selected building blocks provide the appropriate constraint while also transmitting load ($\delta_1 = 4.4°$, $\delta_2 = 3.3°$). CDB building blocks are selected for these subproblems. Note that it was not possible to directly connect in_1 to the output. Instead another point in the design space, P_2, was selected.

Submechanism to connect Input₁ to P₂
Because it was not possible to connect in_1 directly to the output, it is necessary to now connect in_1 to P_2 with a mechanism that satisfies motion requirements. To that end, we calculate the displacement, U_{p2}, results in the desired output displacement, U_{out1}. A new SISO problem results where U_{p2} is the desired *output* displacement, U_{in1}, is the required input displacement, and the intermediate transmission mechanism transmits load between in_1 and P_2. The selected building blocks are shown in Figure 9.20. Note that both input ports exploit mechanism symmetry to provide pure linear stiffness at the input.

Finite element analysis of the resulting mechanism is shown in Figure 9.20 with input₁ actuated and input₂ fixed. Finally, a rapid prototype shows the mechanism with both inputs actuated. The resulting output displacements match the desired U_{out1} and U_{out2} (both translation direction and permitted rotation).

9.5.2.2 Closure

The solution process to address the differential gripping problem was not straight-forward but involved simple decomposition principles. As described at the beginning

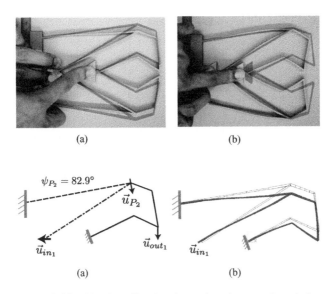

Figure 9.20 Final synthesized mechanism and prototype

of this section, design problems involving multiple ports can be decomposed as input and output constraints and intermediate transmission submechanisms. Design problems may need to be solved in serial and/or parallel, but the general decomposition methodology remains intact.

9.5.3 Displacement Amplifying Mechanisms using Instant Centers (1)

Compliant mechanisms are useful in providing displacement amplification for non-traditional actuators such as MEMS electrostatic and thermal drives and piezostack actuators. Such actuators usually produce high force and low displacement that needs to be amplified for most applications. Many displacement amplifying compliant mechanisms are comprised of C4B and CDB building blocks. Displacements within these mechanisms are well described by instant centers. Instant centers may be utilized to decompose displacement amplification problems.

Figure 9.21 shows the general motion specifications for displacement amplification. Translational directions are specified at the input and output ports. Additionally, the desired geometric advantage, $GA = u_{out}/u_{in}$, is also specified.

A single C4B building block cannot address the displacement amplification problem. By combining two C4B building blocks in series, however, one can solve an arbitrary displacement amplification problem. To do this, one only needs to select a point in the design space where the two C4B building blocks meet. This point is termed the *decomposition point*, DP. There is one translational direction at DP that will yield the desired GA at the output. In Figure 9.21, the translational direction (PCV$_t$) is shown at DP. The translational directions at the input and at DP dictate that the instant center of the first C4B be located at IC$_1$, the intersection of lines perpendicular

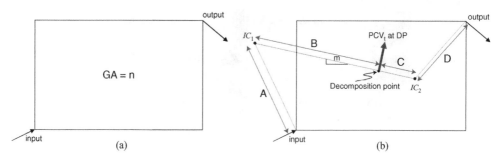

Figure 9.21   Strategy to decompose the displacement amplification problem

to the translational directions through input and DP. IC$_2$ may be found similarly. The translational direction at DP must be selected so that the following equality holds

$$GA = \frac{B}{A}\frac{D}{C}.$$

In general, it is advisable to limit the maximum amplification from a single stage to GA = 3. This is because generating amplification greater than GA = 3 will generate high stress levels or will compromise the mechanism's load bearing capacity. The idea of dividing out the amplification through many stages is similar to using a compound gear train rather than using a single pair of gears to achieve a speed reduction.

9.5.3.1 Design Example: Displacement Inverter

In this example we desire to design a mechanism to invert a vertical displacement and amplify it by a factor of 4 (Figure 9.22a). The DP in Figure 9.22b was selected and the resulting translational direction at the DP is horizontal. The corresponding C4B building blocks are shown in Figure 9.22c. Note that the two building blocks share a common beam, so one of them is eliminated in the final design.

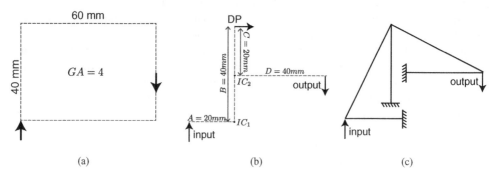

Figure 9.22   (a) Motion-inverting displacement amplification can be accomplished by selecting (b) an appropriate decomposition point and translation direction. (c) The final mechanism satisfies the initial motion specifications

The final design attains the desired geometric advantage and displacement directions within the desired footprint. There are multiple mechanism geometries that would provide the same function if an alternate decomposition point were selected. The decomposition point for the selected mechanism results in an equal distribution of the geometric advantage. A fuller discussion of the selection of the decomposition point may be found in reference [1].

9.6 Conclusions

In this chapter, we presented a building-block approach for the design synthesis of compliant mechanisms. We found that it is possible to design a number of different types of compliant mechanisms using only a small number of building blocks. We utilized various representations of the kinematic function of a compliant mechanism (compliance ellipsoids, eigentwist/eigenwrench parameters, instant centers) and provided methods for decomposing problems into more tractable subproblems.

In sum, the methods presented in this chapter enable a designer to synthesize original compliant mechanisms based only on the desired mechanism behavior. The building-block approach helps to build greater understanding by highlighting the specific function of individual building blocks in an overall design. Such understanding aids in effective mechanism synthesis by including the designer actively in the process. Furthermore, this understanding may be applied to subsequent, distinct problems.

Further Reading

The methods presented in this chapter provide a brief overview of a building-block design approach for compliant mechanisms. The interested reader is encouraged to learn about these methods in greater detail. In addition to the references cited in this chapter (in particular [1–4]), we provide the following list of publications for further study.

Mathematical Modeling of Compliance (including Center of Stiffness)

Lipkin, H., and Patterson, T., 1992, "Geometrical Properties of Modeled Robot Elasticity: Part I—Decomposition," ASME Design Technical Conference and Computers in Engineering Conference, Vol. 45, pp. 179–185.

Lipkin, H., and Patterson, T., 1992, "Geometrical Properties of Modeled Robot Elasticity: Part II—Center of Elasticity," ASME Design Technical Conference and Computers in Engineering Conference, Vol. 45, pp. 187–193.

Loncaric, J., 1985, "Geometrical Analysis of Compliant Mechanisms in Robotics," Ph.D. thesis, Division of Applied Sciences, Harvard University.

Decomposition Methods

Kim, C. J., 2005, "A Conceptual Approach to the Computational Synthesis of Compliant Mechanisms," Ph.D. thesis, University of Michigan.

Krishnan, G., 2010, "An Intrinsic Geometric Framework for the Analysis and Synthesis of Distributed Compliant Mechanisms", Ph.D. thesis, University of Michigan.

Examples of Mechanisms Designed with Building Blocks

Hetrick, J. and Kota, S., 2003, "Displacement amplification structure and device", US Patent 6557436.

Awtar, S., and Slocum, A. H., 2007, "Constraint-Based Design of Parallel Kinematic XY Flexure Mechanisms," *J. Mech. Des.*, **129(8)**, pp. 816–830.

Cappelleri, D. J., Krishnan, G., Kim, C. J., Kumar, V., and Kota, S., 2010, "Toward the Design of a Decoupled, Two-Dimensional, Vision-Based N Force Sensor," *J. Mechanisms Robotics*, **2**, p. 021010. 9 pages.

References

[1] Kim, C., Kota, S., and Moon, Y.-M., "An instant center approach toward the conceptual design of compliant mechanisms," *J. Mechanical Design*, **128**, May 2006, pp. 542–550.

[2] Kim, C., Moon, Y.-M., and Kota, S. "A Building block approach to the conceptual synthesis of compliant mechanisms utilizing compliance and stiffness ellipsoids", *J. Mech. Des.* **130**, 022308 (2008) (11 Pages).

[3] Krishnan, G., Kim, C., and Kota, S., "An intrinsic geometric framework for the building block synthesis of single point compliant mechanisms", *J. Mechanisms Robotics* **3**, 011001 (2011).

[4] Kim, C., "Design strategies for the topology synthesis of dual input-single output compliant mechanisms", *J. Mechanisms Robotics* **1**, 041002 (2009) (10 pages).

[5] Chiou, S.-J. and Kota, S., 1999, "Automated conceptual design of mechanisms," *Mech. Mach. Theory*, **34**, pp. 467–495.

[6] Otto, K. and Wood, K., 2001, *Product Design, Chapter 5: Establishing Product Function*. Prentice Hall, Upper Saddle River, NJ.

Part Four
Library of Compliant Mechanisms

Part Four
Library of Compliant Mechanisms

10

Library Organization

10.1 Introduction

The objective of this chapter[1] is to present the organization and classification scheme of the library section of the handbook. The library has been organized to present compliant elements and designs in a simple and intuitive manner. To achieve this objective, the classification scheme incorporates similar classification techniques used to categorize traditional rigid-body mechanisms, and categorizes mechanisms using a method similar to that employed by Artobolevsky [2]. The scheme classifies mechanisms according to their function, and includes the mechanism depiction and a concise description of its behavior.

10.1.1 Categorization

The purpose of the categorization is to show the reader how the library has been organized in order to efficiently access particular entries, or find entries that achieve particular functions. The classification scheme is divided into different levels, where the classification approach is subdivided into categories, subcategories, classes, then subclasses to appropriately categorize a compliant design. The complete hierarchy of the classification scheme is found in Figure 10.1.

Two systems of categorizing compliant mechanisms were determined to be convenient for engineers: categorizing according to *Elements of Mechanisms* and *Mechanisms*.

Elements of Mechanisms
Compliant *Elements of Mechanisms* are defined as a system of compliant and/or rigid segments that achieve a distinct motion. Understanding the elements used in compliant mechanisms can help engineers understand how a compliant mechanism operates

---

[1] In Proceedings of the ASME IDETC [1]

Handbook of Compliant Mechanisms, First Edition. Edited by Larry L. Howell, Spencer P. Magleby and Brian M. Olsen.
© 2013 John Wiley & Sons, Ltd. Published 2013 by John Wiley & Sons, Ltd.

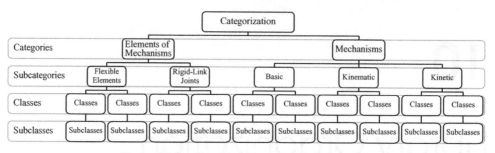

Figure 10.1 Classification scheme hierarchy

and the advantages and disadvantages of these elements. Also, techniques have been established where compliant elements may be used to replace rigid joints [3]. Some examples of elements of compliant mechanisms are the large-displacement elements by Trease et al. [4], the compliant rolling-contact element (CORE) by Cannon and Howell [5], the lamina emergent torsion (LET) joints by Jacobsen et al. [6], and the split-tube flexures by Goldfarb and Speich [7].

The *Elements of Mechanisms* category will be subdivided into two different subcategories, then into different classes where existing designs can be categorized. The two subcategories are: *Flexible Elements* and *Rigid-Link Joints*. It was deemed necessary that the *Rigid-Link Joints* subcategory should be included in this classification because compliant mechanisms utilize both flexible and rigid elements to achieve their kinematic and kinetic behavior. The specific class characterizes the functional operation of the element. In some cases, additional subclasses are appended to a class where there are unique characteristics of elements that needed to be further classified. The subcategories and their subsequent classes for the *Elements of Mechanisms* category are listed in Table 10.1.

Table 10.1 *Elements of Mechanisms'* subcategories and classes

Flexible Elements (FE)	• Beam	(FB)
	• Revolute	(FR)
	– Hinge	(FRH)
	– Scissor	(FRS)
	– Torsion	(FRT)
	– Lamina Emergent	(FRL)
	• Translate	(FT)
	– Lamina Emergent	(FTL)
	• Universal	(FU)
	– Lamina Emergent	(FUL)
	• Flexible Elements: Other	(FO)
Rigid-Link Joints (RLJ)	• Revolute	(RR)
	• Prismatic	(RP)
	• Universal	(RU)
	• Rigid-Link Joints: Other	(RO)

Mechanisms

Mechanisms are defined as a system of rigid bodies connected by elements to achieve a desired motion and/or force transmission. The *Mechanism* category is subdivided into three subcategories: *Kinematic*, *Kinetic*, and *Basic*. Mechanisms with the primary purpose of obtaining a specified motion, path, orientation, or other positioning relationship, are classified under the *Kinematic* subcategory. Those mechanisms with the primary purpose associated with their force–defection relationship, energy storage, or other force- or energy-related function, are classified under the *Kinetic* subcategory. *Basic* mechanisms are those where the kinematics or kinetics of the mechanism is not defined. The motion (kinematics) and force–deflection behavior (kinetics) of compliant mechanisms are highly coupled; however most compliant mechanism applications are designed with a primary function related either to their intended motion or their force–deflection behavior. These subcategories are then subdivided into classes for categorization of existing compliant mechanism designs. Additional subclasses may be appended to a class that will define a unique characteristic of a mechanism, where further classification was required. The subcategories and their subsequent classes for the *Mechanisms* category are listed in Table 10.2.

Limitations of the Classification Scheme

The proposed classification scheme for compliant mechanisms is based upon existing schemes that classify rigid-body mechanisms. As a result, it is difficult to classify compliant elements and mechanisms in a distinct class. This is because (1) mechanisms may by classified in the *Elements of Mechanisms* category because their behavior is similar to the function of a rigid-link element, (2) the mechanisms are not classified by all their kinematic and kinetic characteristics but by their dominating characteristics, and (3) the classification is ever-expanding to accommodate new elements of mechanisms or mechanisms that require a new class in order to be classified.

10.2 Library of Compliant Designs

The organization of the library is described in this section.

Chapters 11, 12, and 13 contain the handbook's library of compliant designs. Associated with each design is a reference number and reference categorization that indicates the subcategory, and class of the design. The reference number specifies the category of the design, followed by an number assigned to the design (i.e EM-# represents an element and M-# represents a mechanism). The first reference categorization specifies the subcategory followed by the second reference categorization that specifies the class. The reference categorization subcategory and class are indicated by indices that can be found in Tables 10.1 and 10.2. The first two letters of the indices indicate the specific class, and if there is a third letter it indicates the specific subclass (e.g. FR is a flexible *revolute* class and FRH is the subclass *hinge* in the flexible *revolute* class).

Table 10.2 *Mechanisms'* subcategories and classes

Basic Mechanism (BA)	• Four-Bar Mechanism	(BF)
	• Six-Bar Mechanism	(BS)
Kinematics (KM)	• Translational	(TS)
	– Precision	(TSP)
	– Large Motion Path	(TSL)
	– Orthogonal	(TSO)
	• Rotational	(RT)
	– Precision	(RTP)
	– Large Motion Path	(RTL)
	– Orthogonal	(RTO)
	• Translation–Rotation	(TR)
	– Precision	(TRP)
	– Large Motion Path	(TRL)
	– Orthogonal	(TRO)
	• Parallel Motion	(PM)
	– Precision	(PMP)
	– Large Motion Path	(PML)
	• Straight Line	(SL)
	• Unique Motion Path	(UP)
	• Stroke Amplification	(SA)
	• Spatial Positioning	(SP)
	– Precision	(SPP)
	• Metamorphic	(MM)
	• Ratchet	(RC)
	• Latch	(LC)
	• Kinematic: Other	(KMO)
Kinetics (KN)	• Energy Storage	(ES)
	– Clamp	(ESC)
	• Stability	(SB)
	– Bistable	(SBB)
	– Multistable	(SBM)
	• Constant Force	(CF)
	• Force Amplification	(FA)
	• Dampening	(DP)
	• Mode	(MD)
	– Buckle	(MDB)
	– Vibration	(MDV)
	• Kinetic: Other	(KNO)

Each design is shown on its own chart to conveniently convey its pertinent information, and to help engineers quickly identify the element or mechanism and its characteristics. Each chart consists of a reference number (indicated in the upper left-hand corner of the chart), name (upper center), reference categorization (upper right-hand corner), drawing (lower left-hand side), description along with any references where more information can be found (right-hand side), and a description of the drawings in an enumerated format (lower right-hand side) for each design, as shown in Figure 10.2.

Element or Mechanism Reference #	Name	SUBCATEGORY CLASS
Figure 1 *(1)*	General description of element or mechanism, and if more information is available a reference will be included.	
Figure 2 *(2)*	1. References the element's or mechanism's segments, and if fabrication is important the figure will be displayed in the manufacturing layout. 2. Displays the element's or mechanism's deformation.	

Figure 10.2 Library of compliant designs template

10.3 Conclusion

Many engineers are not familiar with compliant mechanisms—their function, application, implementation or their advantages. Currently, no library of compliant mechanisms exists with a classification scheme for helping engineers identify potential compliant mechanisms for a design. The purpose of this resource is to increase designer's awareness of compliant mechanisms and help them identify mechanisms appropriate for their applications.

This approach serves as a foundation for creating such a resource. This scheme allows engineers to achieve desired compliant mechanism designs by considering their function and configuration.

References

[1] B.M. Olsen, J.B. Hopkins, L.L. Howell, S.P. Magleby, and M.L. Culpepper, "A proposed extendable classification scheme for compliant mechanisms," in *ASME Design Engineering Technical Conferences, DETC00/DAC*, no. 87290, 2009.
[2] I.I. Artobolevsky, *Mechanisms in Modern Engineering Design: A Handbook for Engineers, Designers, and Inventors*. Moscow: Mir Publishers, 1975, vol 1-5.
[3] M.D. Berglund, S.P. Magleby, and L.L. Howell, "Design rules for selecting and designing compliant mechanisms for rigid-body replacement synthesis," in *ASME Design Engineering Technical Conferences, DETC00/DAC*, 2000.
[4] B.P. Trease, Y. Moon, and S. Kota, "Design of large-displacement compliant joints," *Journal of Mechanical Design, Transactions of the ASME*, vol. 127, no. 4, pp. 788–798, 2005.
[5] J.R. Cannon and L.L. Howell, "A compliant contact-aided revolute joint," *Mechanism and Machine Theory*, vol. 40, no. 11, pp. 1273–1293, 2005.
[6] J.O. Jacobsen, G. Chen, L.L. Howell, and S.P. Magleby, "Lamina emergent torsional (LET) joint," *Mechanism and Machine Theory*, vol. 44, no. 11, pp. 2098–2109, 2009.
[7] M. Goldfarb and J.E. Speich, "A well-behaved revolute flexure joint for compliant mechanism design," *Journal of Mechanical Design*, vol. 121, no. 3, pp. 424–429, 1999.

11

Elements of Mechanisms

11.1 Flexible Elements

11.1.1 Beams

		FE
EM-1	**Fixed-Pinned**	FB

(1)

(2)

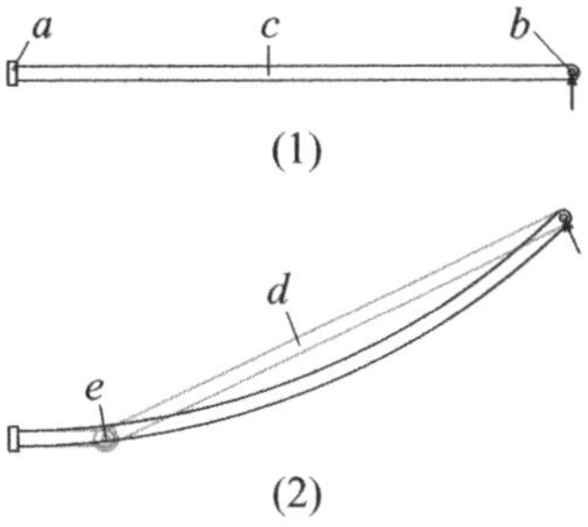

This element is a cantilever beam with a force or moment at the free end. It can be modeled using the pseudo-rigid-body model, which approximates the flexible element as a rigid-link with a torsional spring [1].

(1) Segment a is fixed, segment b is pinned, and segment c is the flexible beam.
(2) Segment c in the deflected position, with its pseudo-rigid-body link d, and torsion spring e.

Handbook of Compliant Mechanisms, First Edition. Edited by Larry L. Howell, Spencer P. Magleby and Brian M. Olsen.
© 2013 John Wiley & Sons, Ltd. Published 2013 by John Wiley & Sons, Ltd.

(1)

(2)

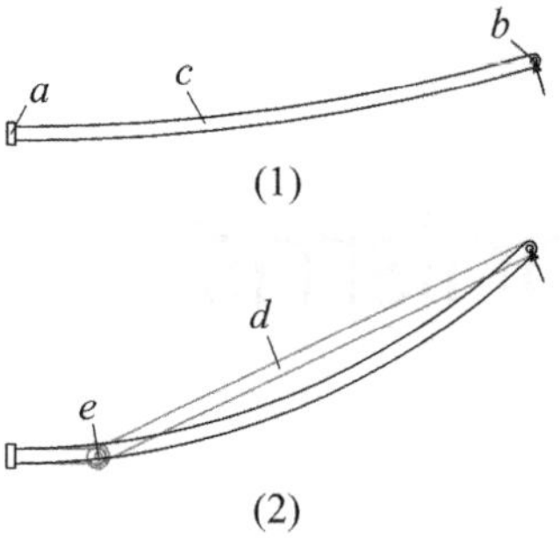

This element is an initially curved cantilever beam with a force or moment at the free end. By using the Bernoulli–Euler equation (curvature is proportional to the moment) a moment can be applied as being an initially-curved beam. This element can be modeled using the pseudo-rigid-body model, which approximates the flexible element as a rigid-link with a torsional spring [1].

(1) Segment a is fixed, segment b is pinned, and segment c is the flexible beam.
(2) Segment c in the deflected position, with its pseudo-rigid-body link d, and torsion spring e.

EM-3   **Fixed-Fixed Compliant Beam**

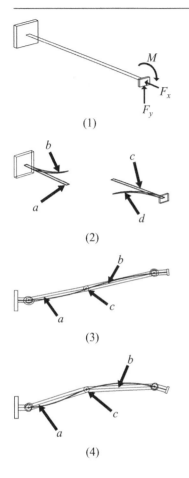

(1)

(2)

(3)

(4)

This element is a fixed-guided compliant segment with varying, specified beam end (guided) angles. Due to the nature of the loading (axial and transverse force, and opposing moment), an inflection point is introduced. Considering that the bending moment at the inflection point is zero, this segment may be treated as two compliant segments separated by the inflection point, each regarded as a fixed-free beam with end forces only and simulated by a pseudo-rigid-body model (PRBM). A variety of combinations of displacement and force boundary conditions may be solved with this modeling approach, including three specified end displacements, vertical end force and moment, two specified end displacements, etc. [1].

(1) This figure shows a fixed-guided compliant beam with end loading.
(2) This figure shows the inflection point used as subdividing the compliant segment, as described above. a and c are the undeformed positions of compliant segments, and b and d the deformed positions, respectively.
(3) This figure shows a case with a positive beam end angle. a and b are the PRBMs of compliant segments in figure 2, and c is regarded as a pin joint assumed at the point of inflection.
(4) This figure shows a case with a negative beam end angle.

| EM-4 | Fixed Guided | FE FB |

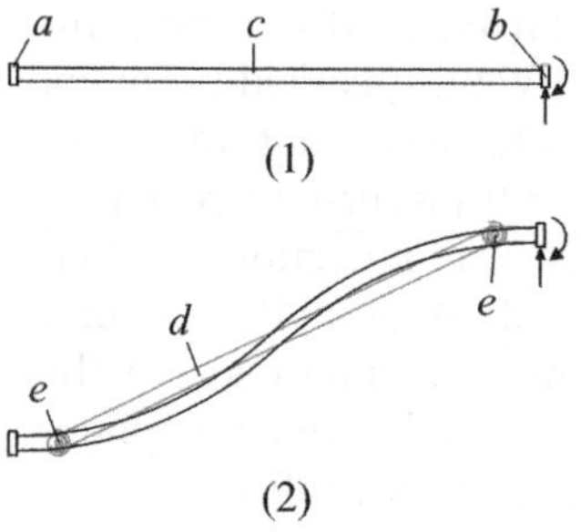

(1)

(2)

This element is a beam fixed at both ends and is a special case for a fixed-fixed case III beam. This occurs when one end goes through a deflection such that the angular deflection at the end remains constant, and the beam shape is antisymmetric about the center. It can be modeled using the pseudo-rigid-body model, which approximates the flexible element as rigid links with torsional springs [1].

(1) Segment *a*, and *b* are fixed, and segment *c* is the flexible beam.
(2) Segment *c* in the deflected position, with its pseudo-rigid-body link *d*, and torsion springs *e*.

<table>
<tr><td></td><td></td><td>FE</td></tr>
<tr><td>**EM-5**</td><td>**Switch Back**</td><td>FB</td></tr>
</table>

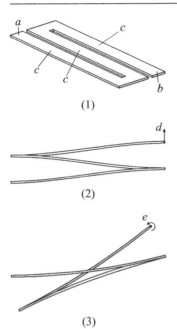

(1)

(2)

(3)

This element is a lamina emergent switch-back flexure because it is fabricated in a plane but has motion outside of the fabrication plane. It is flexible because of the increased length; yet still maintains a compact form. The switch-back can be treated as a fixed-pinned or fixed-fixed element, depending on the boundary conditions [2].

(1) Segments a, and b are attached to a mechanism. Segment c allows flexibility because of its increased length.
(2) Deformed configuration of a fixed-guided deflection in the d direction.
(3) Deformed configuration of a moment e on the end.

<table>
<tr><td></td><td></td><td>FE</td></tr>
<tr><td>**EM-6**</td><td>**Small-Length Flexural Pivot**</td><td>FB</td></tr>
</table>

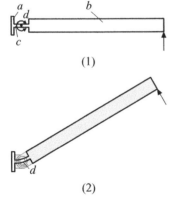

(1)

(2)

This element is a small-length flexural pivot. This element can be approximated as a rigid link and a torsion spring by using the pseudo-rigid-body model, where a general rule is that the length of the flexure is much smaller than the rigid segment length [1].

(1) Rigid segment a, and b are connected by the small-length flexure c. This element rotates about the d-axis.
(2) Deformed configuration of rotation about the d-axis.

EM-7   **L-Shaped Elastic Beam**

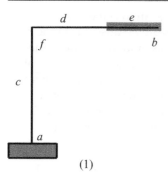

(1)

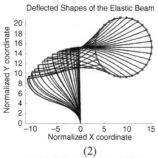

(2)

An L-shaped flexible beam has two flexible beams attached to each other at an angle of 90°. One end of the mechanism is fixed to the ground and the other end is connected to the crank with a free joint. The angle between the two flexible beams doesn't change with the movement of the crank.

(1) Flexible segments c and d have a fixed joint in f with an angle of 90 degrees. At segment a, beam is fixed joint and in segment b, it has a free joint. Segment e is the rotating rigid crank.
(2) Deformed configuration of the flexible beam as the crank rotates.
(3) Prototype of the L-shaped beam.
(4) Prototype of the L-shaped beam.

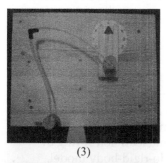

(3)

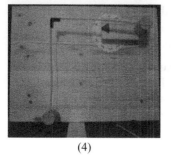

(4)

11.1.2 Revolute

	Compliant Contact-Aided Revolute	FE
EM-8	(CCAR)	FR

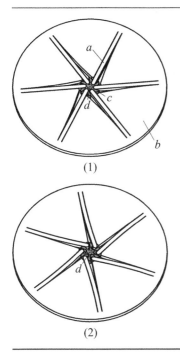

(1)

(2)

This element is a compliant contact-aided revolute (CCAR) joint. It is a planar element capable of performing functions similar to bearings and helical springs. This element can be fabricated at the micro- or macroscale, and can withstand high off-axis loads [3].

(1) Rigid segment b rotates around the rigid segment c about the d-axis. Flexible segments a, provide the energy storage and remain in contact with rigid segment c.

(2) Deformed configuration of rotation about the d-axis.

EM-9   **Multiple-Curve-Beam Flexural Pivot**

(1)

(2)

(3)

This element is a rotational flexural pivot constructed by three curved beams to achieve a large range of motion. Theoretically, this element will rotate without axial-drift motion, because of the symmetric arrangement about the axis [4].

(1) Rigid body *a* is fixed. Rigid body *b* rotates about *c*-axis.
(2) Deformed configuration.
(3) Photo of the device.

Hinge

EM-10	**CORE**	FE/KM FRH/RTL

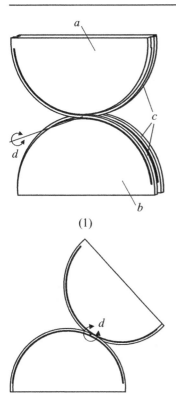

(1)

(2)

This element is a compliant rolling-contact element (CORE) that is designed for compression loads. The CORE connects two rigid links using flexible strips that pass between these rigid-link surfaces, and are attached to the links at the flexure ends. This element is unique such that the axis of rotation changes, which is located at the contact point [3].

(1) Rigid segments *a*, and *b* remain in contact with each other through the flexible segments *c*, while the axis of rotation is at the contact point *d*.
(2) Deformed configuration of rotation about the contact surface.

		FE
EM-11	**Small-Length Flexure**	FRH

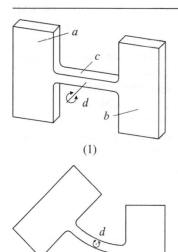

(1)

(2)

This element is a small-length flexural pivot. A small-length flexural pivot is defined as a segment that is significanly shorter and more flexible than its surrounding segments [1].

(1) Rigid segments *a*, and *b* are connected by the flexible segment *c* which rotates about the *d*-axis.
(2) Deformed configuration of rotation about the *d*-axis.

		FE
EM-12	**Living Hinges**	FRH

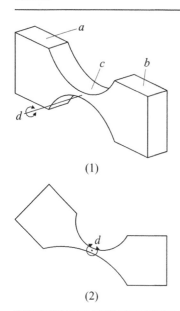

(1)

(2)

This element is a special form of a small-length flexural pivot, where the flexure is extremely short and thin. This element offers little resistance throughout its deflection [1].

(1) Rigid segments *a*, and *b* are connected by the living hinge segment *c* which rotates about the *d*-axis.
(2) Deformed configuration of rotation about the *d*-axis.

FE/KM
FRH/RT

EM-13   **Cross-Axis Flexural Pivot**

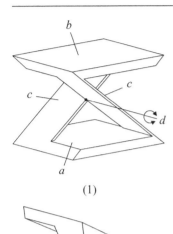

(1)

This element is a cross-axis flexural pivot because it has two flexible beams at an angle. The lengths of the flexible beams are increased because of their angle, but they do not increase the total effective length of the pivot [1, 5].

(1) Rigid segments a, and b are connected by the flexible segments c causing rotation about the d-axis.
(2) Deformed configuration of rotation about the d-axis.

(2)

EM-14   **Statically Balanced Cross-Axis**
                                    Flexural Pivot

FE
FRH

(1)

Cross-flexural pivots are important construction elements in precision engineering. This design adds preloaded double leaf springs that cancel out the rotational stiffness. The result is a zero-stiffness compliant joint [6].

(1) Overview of the design.
(2) Left: cross-flexural pivot; Center: double leaf springs; Right: Assembly.

(2)

EM-15	Constant Stiffness Cross-Axis Flexural Pivot	FE FRH

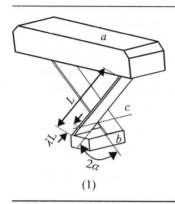

(1)

This element is a rotational flexural pivot with constant stiffness, irrespective of the vertical force applied on the moving stage, when the parameters λ and α satisfy the following condition: $\cos^2 \alpha = \frac{-2(9\lambda^2 - 9\lambda + 1)}{15\lambda}$ [7].

(1) Rigid body a is fixed. Rigid body b rotates about the c-axis.

EM-16	Double Blade Rotary Pivot	FE FRH

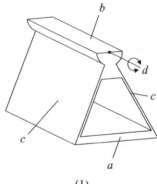

(1)

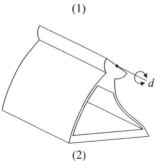

(2)

This element's axis of rotation remains parallel to the plane ground.

(1) Rigid segments a and b are attached to a mechanism. This element achieves compliance by the flexible segments c, causing rotation about the d-axis.
(2) Deformed configuration of rotation about the d-axis.

FE/KN

EM-17   **Bistable Hinge**   FRH/SBB

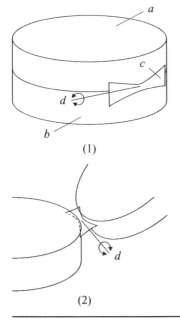

(1)

(2)

This element is a bistable hinge and is designed for applications where there are two desired locations for a link. This element is bistable due the method of the flexible segment attachment to the rigid segment.

(1) Rigid segments a and b are attached to the mechanism. This element achieves compliance by the flexible segment c, causing rotation about the d-axis.
(2) Deformed configuration.

FE

EM-18   **Large Deformation Hinge**   FRH

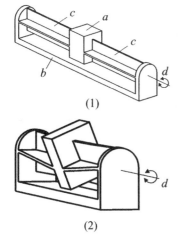

(1)

(2)

This element is designed for large rotations with high off-axis stiffness. The geometry of the cross-plates allow a high degree of flexibility in torsion [8].

(1) Rigid segments a and b are attached to the two cross-plates c that rotate about the d-axis.
(2) Deformed configuration of rotation about the d-axis.

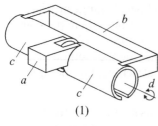

(1)

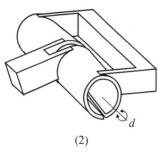

(2)

This element is designed for large rotations with high off-axis stiffness. The split-tube flexures rely on torsion for their flexibility [9].

(1) Rigid segments a and b are attached to the split-tube flexures c which rotate about the d-axis.
(2) Deformed configuration of rotation about the d-axis.

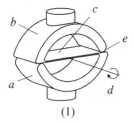

(1)

This element is an isolated-based high-compression compliant mechanism (HCCM) that is designed for rotational motion with the mechanism in compression [10].

(1) Rigid bodies a and b rotate about the d-axis. Segment c is the compliant segment and segment e remains in contact with rigid body b.

		FE/KM
EM-21	**Isolation-based HCCM**	FRH/RTL

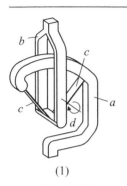

(1)

This mechanism is an inversion-based high-compression compliant mechanism (HCCM) that is designed for rotational motion with the mechanism in compression [10].

(1) Rigid bodies a and b rotate about the d-axis, by flexible segments c.

		FE/KM
EM-22	**Isolation-based HCCM**	FRH/RTL

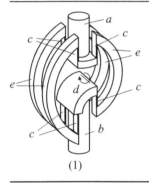

(1)

This element is an inversion-based high-compression compliant mechanism (HCCM) that is designed for rotational motion with the mechanism in compression [11].

(1) Rigid bodies a and b rotate about the d-axis, by flexible segments c that is connected by rigid body e.

		FE/KM
EM-23	HCCM	FRH/RTL

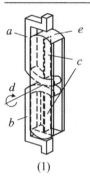

(1)

This element is an inversion-based high-compression compliant mechanism (HCCM) that is designed for rotational motion with the mechanism in compression [11].

(1) Rigid bodies a and b rotate about the d-axis which is where both rigid bodies come in contact. Flexible segments c allow the rotation and are connected by rigid body e.

	Rotational Joint with Fixed Rotational	FE
EM-24	Center and Locking System	FRH

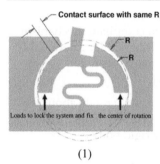

(1)

(2)

The rotational compliant joint with a fixed center of rotation and locking system. Once the screws are engaged, the spring will be stretched and the outer side of the joint will be pushed against the inner side of its frame that is of the same radius for the contact area, so the system will be locked and the center of rotation remains the same [12].

(1) Concept of rotational joint with fixed rotational center and locking system.
(2) Prototype in deformed position and locking system activated.

EM-25	**ADLIF: A Large-Displacement** **Beam-Based Flexure Joint**	FE FRH

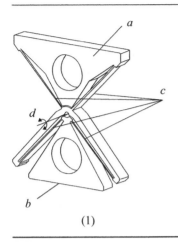

(1)

This element is an anti-symmetric double leaf-type isosceles-trapezoidal flexure joint (ADLIF), which is constructed by four identical beams and the extended lines of the four leaves intersected at a virtual pivot point [13].

(1) Rigid segments a and b are attached to the leaf-type flexures c that rotate about the d-axis.

EM-26   **Curve-Beam Flexural Pivot**

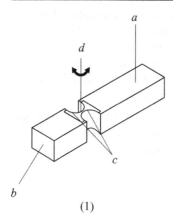

(1)

Intesifying Tendon

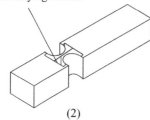

(2)

Coupler

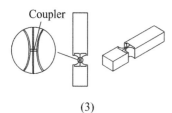

(3)

This element is a rotational flexural pivot constructed by two curved beams to achieve a large range of motion. The geometry is derived from the conventional notch-type flexural pivots. The main feature is that the cross section of the pivot is different from those of conventional pivots: the front pivot has a hollow section with two curved leaves but the rear has a solid section with two curved edges [14].

(1) Rigid segments a and b are attached to the leaf-type flexures c that rotate about the d-axis.
(2) One derived configuration with an intensifying tendon.
(3) One derived configuration with a coupler.

Scissor

| EM-27 | Deltoid Q-Joint | FE FRS |

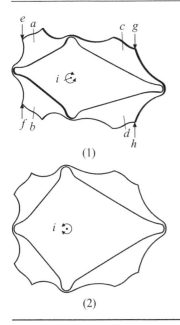

(1)

(2)

This element is a deltoid-type Q-joint. It is constructed when each rigid segment in the quadrilateral is made adjacent to a segment of equal length [1].

(1) Rigid segments a and b, and rigid segments c and d, are equal length, respectively. When the rigid segments a and b deform in the e-, and f-directions, respectively, the rigid segments c and d deform in the g- and h-directions, respectively. This element appears to rotate about the i-axis.

(2) Deformed configuration. The angle between rigid segments a, and b, and c, and d, respectively, decreases.

| EM-28 | Scissor Hinge | FE FRS |

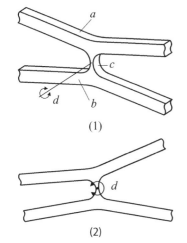

(1)

(2)

This element allows a scissor action by use of a small flexure placed in the middle of rigid segments.

(1) Rigid segments a and b are connected by the flexible segment c and rotate about the d-axis.

(2) Deformed configuration of rotation about the d-axis.

Torsion

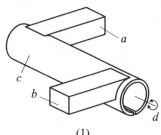

This element is a split-tube flexure. It is compliant in the desired axis of rotation but stiff in its other axes [9].

(1) Rigid segments a and b are attached to the split-tube flexure c which rotates about the d-axis.

(2) Deformed configuration of rotation about the d-axis.

(1)

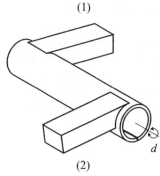

(2)

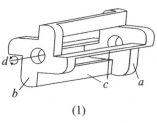

This element uses beams in a circular array allowing rotation. When the rotation is large, the length of the element retracts.

(1) Rigid segments a and b are attached to a mechanism. Flexible segments c allow rotation about the d-axis.

(2) Deformed configuration of rotation about the d-axis.

(1)

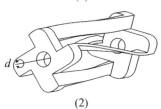

(2)

FE
EM-31   **Torsion Translator**   FRT

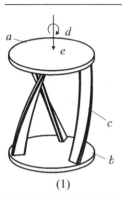

(1)

This element uses beams in a circular array allowing rotation. When the rotation is large, the length of the element retracts.

(1) Rigid segments *a* and *b* are connected by the flexible segments *c* causing rotation about the *d*-axis. If element undergoes a large rotation, it will translate in the *e*-direction.

FE
EM-32   **Tubular Cross-Axis Flexural Pivot**   FRT

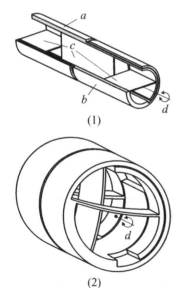

(1)

(2)

This element is a cross-axis flexural pivot because it has two flexible beams at an angle. The lengths of the flexible beams are increased because of their angle, but they do not increase the total effective length of the pivot [1, 5].

(1) Rigid segments *a* and *b* are connected by the flexible segments *c* causing rotation about the *d*-axis.
(2) Deformed configuration of rotation about the *d*-axis.

Lamina Emergent

		FE
EM-33	**Reduced Inside Area Joint**	FRL

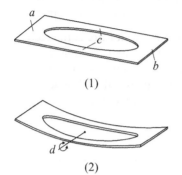

This element's inside area is reduced, allowing greater flexibility. It is suited for applications where angular rotation is desired. This element can also be fabricated in a single plane (lamina emergent) [2].

(1) Rigid segments a and b are attached to a mechanism. Segments c are flexible compared to the rest of the segments because of the reduced cross-sectional area, which allows rotation about the d-axis.

(2) Deformed configuration of rotation about the d-axis.

		FE
EM-34	**Reduced Outside Area Joint**	FRL

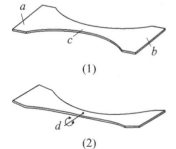

This element's outside areas are reduced, allowing greater flexibility. It is suited for applications where angular rotation is desired. This element can be fabricated in a single plane (lamina emergent) [2].

(1) Rigid segments a and b are attached to a mechanism. Segment c is flexible compared to the rest of the segments because of the reduced cross-sectional area, which allows rotation about the d-axis.

(2) Deformed configuration of rotation about the d-axis.

FE
EM-35   **Outside LET Joint**   FRL

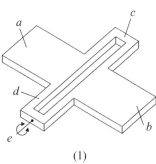

(1)

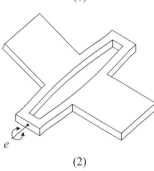

(2)

This element is a lamina emergent torsion (LET) joint. It is suited for applications where large angular rotation is desired, but high off-axis stiffness is not critical. It can be fabricated in a single plane [15].

(1) Rigid segments a and b are attached to a mechanism. The flexible segments c and d are in bending and torsion, respectively, causing a rotation about the e-axis.

(2) Deformed configuration of rotation about the e-axis.

		FE
EM-36	**Inside LET Joint**	FRL

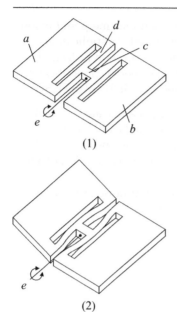

(1)

(2)

This element is a lamina emergent torsion (LET) joint. It is suited for applications where large angular rotation is desired but high off-axis stiffness is not critical. It can be fabricated in a single plane [15].

(1) Rigid segments *a* and *b* are attached to a mechanism. The flexible segments *c* and *d* are in bending and torsion, respectively, causing a rotation about the *e*-axis.
(2) Deformed configuration of rotation about the *e*-axis.

		FE
EM-37	**Notch Joint**	FRL

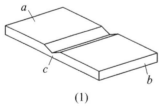

(1)

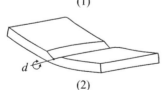

(2)

This element is designed for angular rotation and can be fabricated in a single plane (lamina emergent). The reduced thickness of this element allows for greater flexibility [2].

(1) Rigid segments *a* and *b* are attached to a mechanism. Segment *c* allows flexibility because of its reduced thickness, enabling rotation about the *d*-axis.
(2) Deformed configuration of rotation about the *d*-axis.

	FE
EM-38	**Groove Joint**	

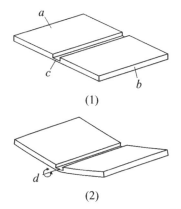

This is a lamina emergent groove joint. It is suited for applications where angular rotation is desired. The reduced thickness allows greater flexibility [2].

(1) Rigid segments *a* and *b* are attached to a mechanism. Segment *c* allows flexibility because of its reduced thickness, enabling rotation about the *d*-axis.

(2) Deformed configuration of rotation about the *d*-axis.

11.1.3 Translate

	FE
EM-39	**Leaf Spring Translational Joint**	FT

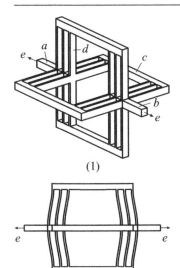

This element is a leaf spring translational joint that is designed to have high off-axis stiffness. This element has a relatively large range of motion [8].

(1) Rigid segments *a* and *b* are attached to a mechanism. If the rigid segment *a* is fixed then the rigid segment *b* translates in the *e*-direction. Segments *c* are rigid and segments *d* are flexible.

(2) Deformed configuration of translation in the *e*-direction.

		FE
EM-40	**Two Force Member**	FT

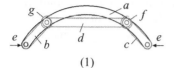

(1)

This element is a beam that is pinned on both ends (two-force member). It is initially curved in mode one buckling. This simple element can be modeled using the pseudo-rigid-body model, which approximates the flexible elements as a rigid link with a torsional spring [1].

(1) The deflected position of the flexible beam a, with segments b, c, and d as the pseudo-rigid-body link, deflected in the e-direction with torsional springs f and g.

Lamina Emergent

		FE
EM-41	**LEM Translator**	FTL

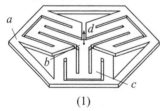

(1)

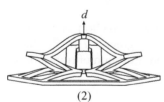

(2)

This element is a lamina emergent translator. It utilizes switch-back beams for a compact translational motion. This element can be fabricated in a single plane.

(1) Rigid segments a and b are attached to a mechanism. The flexible segments c are switch back beams that allow flexibility, which is able to translate in the d-direction

(2) Deformed configuration of translation in the d-direction.

11.1.4 Universal

EM-42	Ortho Skew Double Rotary	FE/KM FU/RTO

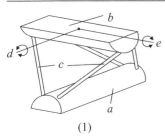

(1)

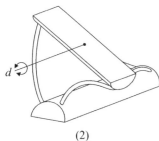

(2)

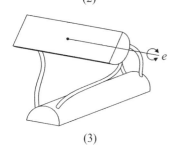

(3)

This element's axes of each of the four constraints intersect both lines of rotation.

(1) Rigid segments a and b are attached to a mechanism. The flexible constraints c allow rotation about the d- and e-axes.
(2) Deformed configuration of rotation about the d-axis.
(3) Deformed configuration of rotation about the e-axis.

EM-43	Tripod Spherical Joint	FE/KM
		FU/RTO

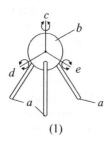

(1)

(2)

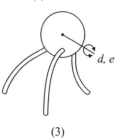

(3)

This element has three orthogonal rotational degrees of freedom. This flexure emulates the degrees of freedom of a spherical ball joint.

(1) Segments *a* are attached to a mechanism. Rigid segment *b* is able to rotate about the *c*-, *d*-, and *e*-axes, by the flexible constraints *a*.

(2) Deformed configuration of rotation about the *c*-axis.

(3) Deformed configuration of rotation about the *d*- or *e*-axis.

Lamina Emergent

		FE
EM-44	**Reduced Outside Area Joint**	FUL

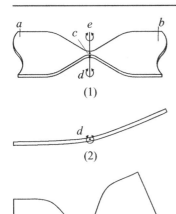

(1)

(2)

(3)

This element is a unique outside reduced area joint, such that the width of the reduced area is similar to its thickness. This reduces the off-axis stiffness and the rotational element becomes a universal element. This element can be fabricated in a single plane [2].

(1) Rigid segments *a* and *b* are attached to a mechanism. Segment *c* is flexible compared to the rest of the segments because of the reduced cross-sectional area, which allows rotation about the *d*- and *e*-axes.

(2) Deformed configuration of rotation about the *d*-axis.

(3) Deformed configuration of rotation about the *e*-axis.

EM-45   **Outside LET Joint**

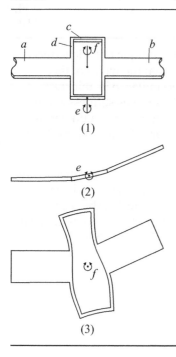

(1)

(2)

(3)

This element is a unique inside lamina emergent torsion (LET) joint where the torsional hinges are on the outside of the element. In this case, the off-axis stiffness is reduced, allowing the rotational element to become a universal element. This element can be fabricated in a single plane [15].

(1) Rigid segments *a* and *b* are attached to a mechanism. The flexible segments *c* and *d* are in bending and torsion, respectively, causing a rotation about the *e*- and *f*-axes.
(2) Deformed configuration of rotation about the *e*-axis.
(3) Deformed configuration of rotation about the *f*-axis.

EM-46   **Inside LET Joint**

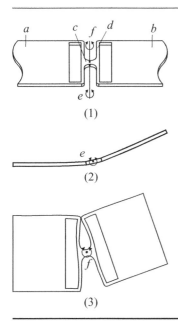

(1)

(2)

(3)

This element is a unique inside lamina emergent torsion (LET) joint where the torsional hinges are on the inside of the element. In this case, the off-axis stiffness is reduced, allowing the rotational element to become a universal element. This element can be fabricated in a single plane [15].

(1) Rigid segments a and b are attached to a mechanism. The flexible segments c and d are in bending and torsion, respectively, causing a rotation about the e- and f-axes.
(2) Deformed configuration of rotation about the e-axis.
(3) Deformed configuration of rotation about the f-axis.

11.2 Rigid-Link Joints

11.2.1 Revolute

		RLJ
EM-47	**Revolute Joint**	RR

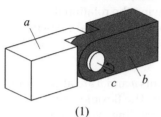

(1)

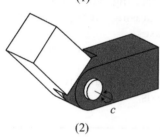

(2)

This element is a lower kinematic pair that provides one rotational degree of freedom between connected links [2].

(1) Rigid segments a and b rotate about the c-axis.
(2) Deformed configuration of rotation about the c-axis.

EM-48	Passive Joint	RLJ RR

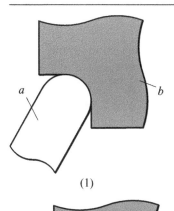

(1)

(2)

This element allows rotation between two rigid segments without using a traditional pin joint. These segments need to be in contact to operate [1].

(1) Rigid segments a and b are to remain in contact, allowing rotation about the c-axis.
(2) Deformed configuration of rotation about the c-axis.

11.2.2 Prismatic

EM-49	Prismatic Joint	RLJ RP

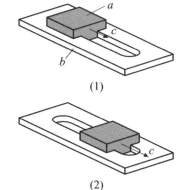

(1)

(2)

This element is a lower kinematic pair that provides one translational degree of freedom between connected links [2].

(1) Rigid segment a translates in the c-direction on rigid segment b.
(2) Deformed configuration of translation in the c-direction.

11.2.3 Universal

EM-50	Universal Joint	RLJ RU

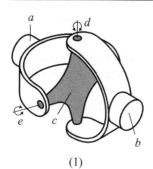

(1)

(2)

This element provides two rotational degrees of freedom between connected links [2].

(1) Rigid segments a and b rotate about the d- and e-axes, by rigid segment c.
(2) Deformed configuration of rotation about the d- and e-axes.

11.2.4 Others

EM-51	Half Joint	RLJ RO

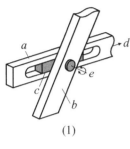

(1)

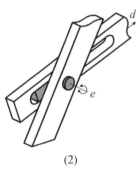

(2)

This element provides one rotational and one translational degree of freedom between connected links. The axis of rotation is orthogonal to the direction of translation [2].

(1) Rigid segment a translates in the d-direction and rotates about the e-axis on rigid segement c, which is attached to ground b.
(2) Deformed configuration of translation in the d-direction and rotation about the e-axis.

EM-52	Spherical Joint	RLJ RO

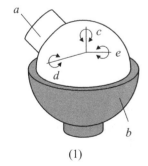

(1)

This element is a lower kinematic pair that provides three rotational degrees of freedom between connected links [2].

(1) Rigid segments a and b rotate about the c-, d-, and e-axes.

EM-53	Planar Joint	RLJ RO

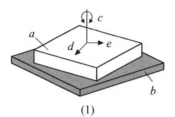

(1)

This element is a lower kinematic pair that provides two translational and one rotational degrees of freedom between connected links. The two translation directions are coplanar and the axis of rotation is orthogonal to that plane [2].

(1) Rigid segment *a* translates in the *d*- and *e*-directions and rotates about the *c*-axis on rigid segment *b*, which is attached to ground.

EM-54	Helical Joint	RLJ RO

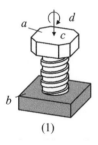

(1)

This element is a lower kinematic pair that provides both translation and rotation between connected links. The rotation and translation are coupled in such a way as to allow a single degree of freedom. The axis of rotation and the direction of translation are collinear [2].

(1) Rigid segment *a* translates in the *c*-direction and rotates about the *d*-axis on rigid segment *b*, which is attached to ground.

EM-55	Cylindric Joint	FE RO

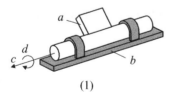

(1)

This element is a lower kinematic pair that provides both translation and rotation between connected links. The rotation and translation are not coupled, so this joint has two degrees of freedom. The axis of rotation and the direction of translation are collinear [2].

(1) Rigid segment *a* translates in the *c*-direction and rotates about the *d*-axis on rigid segment *b*, which is attached to ground.

References

[1] L. L. Howell, *Compliant Mechanisms*. New York, NY, Wiley-Interscience, July 2001.

[2] B. G. Winder, S. P. Magleby, and L. L. Howell, "A study of joints suitable for lamina emergent mechanisms," in *ASME Design Engineering Technical Conferences, DETC08*, 2008.

[3] J. R. Cannon and L. L. Howell, "A compliant contact-aided revolute joint," *Mechanism and Machine Theory*, vol. 40, no. 11, pp. 1273–1293, 2005.

[4] S. S. Zhao, S. S. Bi, J. J. Yu, and X. Pei, "Curved beam flexure element based of large deflection annulus figure flexure hinge," in *ASME Design Engineering Technical Conferences, DETC08/DAC*, 2008.

[5] B. D. Jensen and L. L. Howell, "The modeling of cross-axis flexural pivots," *Mechanism and Machine Theory*, vol. 37, no. 5, pp. 461–476, 2002.

[6] F. Morsch and J. L. Herder, "Design of a generic zero stiffness compliant joint," in *ASME Design Engineering Technical Conferences, DETC10/DAC*, 2010.

[7] H. Z. Zhao and S. S. Bi, "Stiffness and stress characteristics of the generalized cross-spring pivot," *Mechanisms and Theory*, vol. 45, no. 3, pp. 378–391, 2010.

[8] B. P. Trease, Y. Moon, and S. Kota, "Design of large-displacement compliant joints," *Journal of Mechanical Design, Transactions of the ASME*, vol. 127, no. 4, pp. 788–798, 2005.

[9] M. Goldfarb and J. E. Speich, "A well-behaved revolute flexure joint for compliant mechanism design," *Journal of Mechanical Design*, vol. 121, no. 3, pp. 424–429, 1999.

[10] A. E. Guérinot, S. P. Magleby, L. L. Howell, and R. H. Todd, "Compliant joint design principles for high compressive load situations," *Journal of Mechanical Design*, vol. 127, p. 774, 2005.

[11] A. E. Guerinot, S. P. Magleby, and L. L. Howell, "Preliminary design concepts for compliant mechanism prosthetic knee joints," in *ASME Design Engineering Technical Conferences, DETC04/DAC*, vol. 2 B, 2004, pp. 1103–1111.

[12] L. Kluit, N. Tolou, and J. L. Herder, "Design of a tunable fully compliant stiffness comensation mechanism for body powered hand prosthesis," in *In proceeding of the Second International Symposium on Compliant Mechanisms CoMe2011*, 2011.

[13] X. Pei and J. J. Yu, "A new large-displacement beam-based flexure joint," *Mechanical Sciences*, Vol. 2 pp. 183–188, 2011.

[14] J. J. Yu, G. H. Zong, and S. S. Bi, "A new family of large-displacement flexural pivots," in *ASME Design Engineering Technical Conferences, DETC07/DAC*, 2007.

[15] J. O. Jacobsen, "Fundamental components for lamina emergent mechanisms," Master's thesis, Brigham Young University, Dept. of Mechanical Engineering, 2008.

12

Mechanisms

12.1 Basic Mechanisms

12.1.1 Four-Bar Mechanism

M-1	Compliant Mechanism Type Synthesis	BA BF

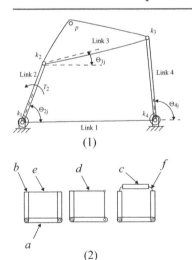

A compliant mechanism type synthesis approach, which is heuristic in nature, begins with treating a rigid-body four-bar mechanism with a torsional spring located at one or more (up to four) of the kinematic pairs. Each of these combinations, regarded as a pseudo-rigid-body model (PRBM), will correspond to multiple equivalent compliant mechanisms. This formulation embraces the methodology in which compliant mechanisms are represented by PRBMs [1, 2].

(1) This figure shows the initial pseudo-rigid-body model (PRBM) that may have a torsional spring at each revolute pair.

Handbook of Compliant Mechanisms, First Edition. Edited by Larry L. Howell, Spencer P. Magleby and Brian M. Olsen.
© 2013 John Wiley & Sons, Ltd. Published 2013 by John Wiley & Sons, Ltd.

BA	
M-1 **Compliant Mechanism Type Synthesis**	BF

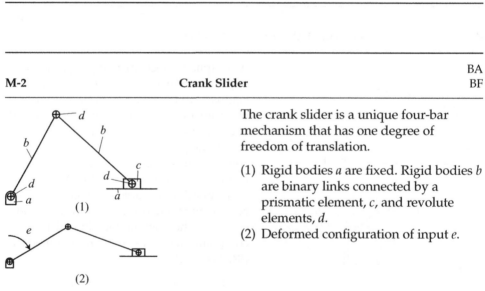

(2) This figure exemplifies the typical elements used in arriving at the slate of possible compliant mechanisms. *a* - Rigid Pinned-Pinned Segment, *b* - Rigid Fixed-Pinned Segment, *c* - Rigid Fixed-Fixed Segment, *d* - Compliant Fixed-Pinned Segment, *e* - Compliant Fixed-Fixed Segment, and *f* - Small-Length Flexural Pivot.

(3) Using the fundamental elements described in (2), 18 resulting compliant mechanism configurations are shown in this figure.

(3)

BA	
M-2 **Crank Slider**	BF

The crank slider is a unique four-bar mechanism that has one degree of freedom of translation.

(1) Rigid bodies *a* are fixed. Rigid bodies *b* are binary links connected by a prismatic element, *c*, and revolute elements, *d*.

(2) Deformed configuration of input *e*.

(1)

(2)

12.1.2 Six-Bar Mechanism

M-3   **Watt Inversion I**

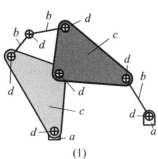

(1)

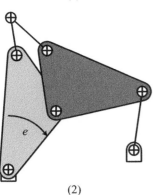

(2)

A Watt Mechanism is a six-bar mechanism characterized as having its two ternary links connected. This mechanism is in the inversion I configurations.

(1) Rigid bodies a are fixed. Rigid body binary, b, and ternary, c, links are connected by revolute elements, d.
(2) Deformed configuration of input e.

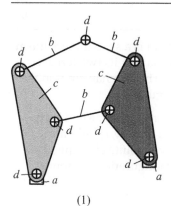

(1)

A Stephenson Mechanism is a six-bar mechanism characterized as having its two ternary links separated by a binary link. This mechanism is in the inversion I configurations.

(1) Rigid bodies *a* are fixed. Rigid body binary, *b*, and ternary, *c*, links are connected by revolute elements, *d*.
(2) Deformed configuration of input *e*.

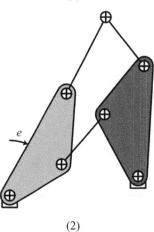

(2)

12.2 Kinematics

12.2.1 Translational

		KM
M-5	**X Bob**	TS

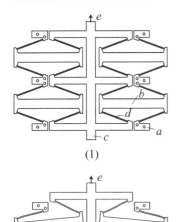

(1)

(2)

This mechanism is a fully compliant linear-motion mechanism with high off-axis stiffness. The design is based on a multiple Roberts four-bar approximate straight-line mechanism and by using symmetry [3].

(1) Rigid bodies a are fixed. Segments b are rigid. Rigid body c translates in the e-direction by flexible segments d.
(2) Deformed configuration of translation in the e-direction.

		KM/KN
M-6	**SRFBM**	TS/SBB

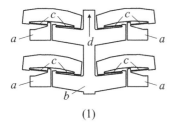

(1)

This mechanism is a fully compliant bistable mechanism that has been developed for applications in microswitching. The application is a self-retracting fully compliant bistable micromechanism (SRFBM) [4].

(1) Rigid bodies a are fixed. Rigid body b translates in the d-direction. Flexible segments c provide the bistable and translating motion.

M-7 | **Bistable Planar Translator** |

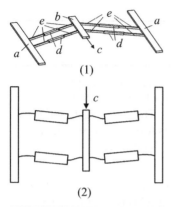

(1)

(2)

This mechanism has a planar-linear translating motion. This mechanism is also bistable [5].

(1) Rigid bodies *a* are fixed. Rigid body *b* translates in the *c*-direction. Flexible segments *e* are separated by a rigid segment *d*, which provides the bistable and translating motion.
(2) Deformed configuration of translation in the *c*-direction.

M-8 | **Translational Bistable Planar** |

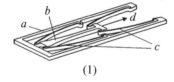

(1)

This mechanism has a planar-linear translating motion. This mechanism is also bistable.

(1) Rigid bodies *a* are fixed. Rigid body *b* translates in the *d*-direction. Flexible segments *c* provide the bistable and translating motion.

M-9 | **Parallel Bistable Translator** |

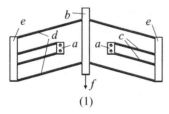

(1)

This mechanism has mirrored and folded parallel-guiding mechanisms. It allows a translational degree of freedom with high off-axis stiffness. Also, the flexures are fabricated at an offset angle, which allows the bistability [6].

(1) Rigid bodies *a* are fixed. Rigid body *b* translates in the *f*-direction. Flexible segments *c* and *d* are separated by a rigid segment *e*, which provides the bistable and translating motion.

M-10	Zero-Force or Bistable Translation Mechanism	KM/KN TS/SBB

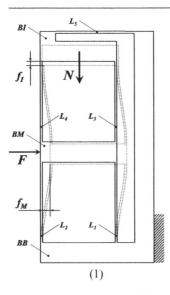

(1)

(2)

(3)

A preloaded spring (L_5) is used in combination with a parallel spring stage (L_1 & L_2) in an arrangement where it loses elastic energy while the flexure is moved away from its neutral position. This leads to an almost constant global spring constant, i.e. the total amount of elastic energy stored in the structure is independent of the position of the mobile block BM. The preload force can be tuned in order to approach a zero-force (and zero-stiffness) over the full motion range, or to produce a bistable behavior.

(1) A constant preload force N is applied to the intermediate block BI by an external spring or by the blade L_5. This force reduces, cancels or inverts the sign of the stiffness of the four blades (L_1 to L_4).

(2) Photograph of a mock-up with an external preload spring.

(3) Photograph of a monolithic zero-force translation device (except for the two dark shims used to preload the thick vertical blade) manufactured by EDM.

		KM
M-11	**Parallel Translator**	TS

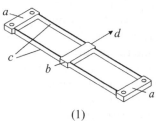

(1)

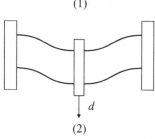

(2)

This mechanism has mirrored parallel-guiding mechanisms. This mechanisms allows a translational degree of freedom with high off-axis stiffness [7].

(1) Rigid bodies *a* are fixed. Rigid body *b* translates in the *d*-direction. Flexible segments *c* allow the degree of translation.
(2) Deformed configuration of translation in the *d*-direction.

		KM
M-12	**Collinear-Type Statically Balanced Linear Motion**	TS

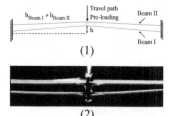

(1)

(2)

This mechanism is a zero-stiffness mechanism with a near-zero actuation force in a finite range of motion due to cooperative action of bistable beams with different shapes. $h_{\text{Beam I}} \neq h_{\text{Beam II}}$ [8]

(1) Concept of collinear-type statically balanced linear motion compliant mechanisms; $h_{\text{Beam I}}$ and $h_{\text{Beam II}}$ are corresponding to rises, h, of Beam I and Beam II, respectively.
(2) Prototype (30–1000 μm) and deformed shape.

KM
M-13	**1DOF Translation Mechanism**	TS

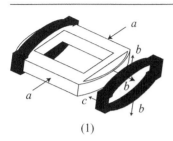

(1)

The translation mechanism allows for one degree of translation. The other degrees of freedom have been cancelled out due to the moment of inertia of the leaf springs [9].

(1) The deformation in leaf springs *b* allows for only translation in the *a*-direction. The *c* arrows indicate the mechanism's assembly scheme.

	Statically Balanced Compliant	KM
M-14	**Laparoscopic Grasper**	TS

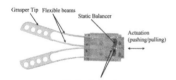

(1)

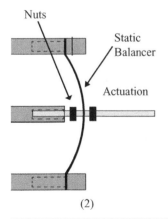

(2)

The mechanism is a zero-stiffness mechanism with a near-zero actuation force for a finite range of motion. The positive stiffness of the grasper has been compensated by bistable straight guided beams preloaded collinear to the motion path of the grasper [10, 11, 12].

(1) Concept of the statically balanced compliant laparoscopic grasper.
(2) Tuning the negative stiffness.

Precision

	KM
M-15 **Precision Cross-Bladed Translator**	TSP

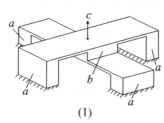

(1)

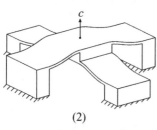

(2)

This mechanism is suited for precision applications where only one translational degree of freedom is required. All other motions are constrained. The translational degree of freedom is orthogonal to the plane of the ground [13].

(1) Rigid bodies *a* are fixed. Rigid body *b* is free to translate in the *c*-direction.
(2) Deformed configuration when translating in the *c*-direction.

	KM
M-16 **Parallel Blade Translator**	TSP

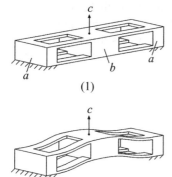

(1)

(2)

This mechanism is suited for precision applications where only one translational degree of freedom is required. All other motions are constrained. The translational degree of freedom is orthogonal to the ground plane [13].

(1) Rigid bodies *a* are fixed. Rigid body *b* is free to translate in the *c*-direction.
(2) Deformed configuration when translating in the *c*-direction.

M-17   **End Effector**

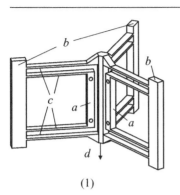

(1)

This mechanism has three folded parallel-guiding mechanisms that allow a translational degree of freedom with high off-axis stiffness [14].

(1) Rigid bodies *a* are fixed. Rigid body *b* translates in the *d*-direction. Flexible Segments *c* allow the degree of translation.

Large Motion Path

M-18   **Parallel Translator**

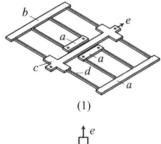

(1)

This mechanism has mirrored and folded parallel-guiding mechanisms. This mechanism allows a translational degree of freedom with high off-axis stiffness. This mechanism can be used in microelectromechanical systems (MEMS). It is often referred to as a "folded beam suspension."

(1) Rigid bodies *a* are fixed. Segments *b* are rigid. Rigid body *c* translates in the *e*-direction by flexible segments *d*.
(2) Deformed configuration of translation in the *e*-direction.

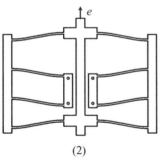

(2)

Orthoginal

	Perpendicular-Type Statically	KM
M-19	**Balanced Linear Motion**	TSO

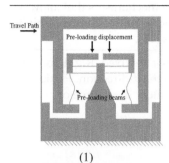

(1)

(2)

This mechanism is a zero-stiffness mechanism with a near-zero actuation force in a finite range of motion due to preloading perpendicular to the motion of the mechanism [8].

(1) Concept of perpendicular-type statically balanced linear motion compliant mechanisms.
(2) Prototype (30–1000 μm) and deformed shape.

12.2.2 Rotational

See also

Name	Reference Index	Categorization Index
Cross-Axis Flexural Pivot	EM-13	FE/KM
		FRH/RT

Precision

<table>
<tr><td></td><td></td><td>KM</td></tr>
<tr><td>M-20</td><td>**Offset Parallel Translator**</td><td>TSP</td></tr>
</table>

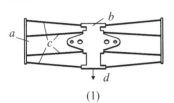

(1)

This mechanism has mirrored parallel-guiding mechanisms that allow a translational degree of freedom with high off-axis stiffness. Also, due to the offset of the flexures, the mechanism doesn't allow a high degree of motion, but has repeatable motion.

(1) Rigid bodies *a* are fixed. Rigid segment *b* translates in the *d*-direction by flexible segments *c*.

<table>
<tr><td></td><td></td><td>KM</td></tr>
<tr><td>M-21</td><td>**Precision Constraint Rotator**</td><td>RTP</td></tr>
</table>

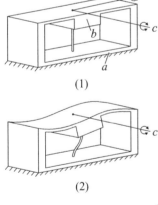

(1)

(2)

This mechanism is suited for precision applications where two orthogonal rotational degrees of freedom are required. The orthogonal rotational degrees of freedom are parallel to the plane of the ground [13].

(1) Rigid body *a* is fixed. Rigid body *b* rotates about the *c*- and *d*-axes.
(2) Deformed configuration of rotation about the *c*-axis.
(3) Deformed configuration of rotation about the *d*-axis.

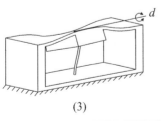

(3)

		KM
M-22	**Precision Constraint Rotator**	RTP

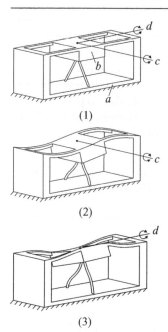

(1)

(2)

(3)

This mechanism is suited for precision applications where two orthogonal rotational degrees of freedom are required. The rotational degrees of freedom are parallel to the plane of the ground [13].

(1) Rigid body a is fixed. Rigid body b rotates about the c- and d-axes.
(2) Deformed configuration of rotation about the c-axis.
(3) Deformed configuration of rotation about the d-axis.

Large Motion Path

		KM
M-23	**Rotational LEM**	RTL

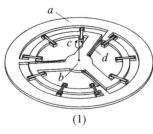

(1)

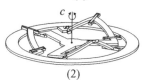

(2)

This is a spherical lamina emergent mechanism (LEM) that emerges out of the manufactured plane when rotated [15].

(1) Rigid body a is fixed. Rigid body b rotates about the c-axis.
(2) Deformed configuration of rotation about the c-axis.

		KM
M-24	Bricard 6R (LEM)	RTL

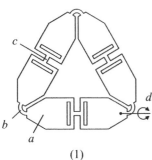

(1)

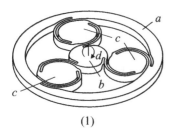

(2)

This is a Bricard 6R fully compliant lamina emergent mechanism (LEM). This mechanism allows infinite rotation [16, 17].

(1) Rigid body a rotates by small length flexure b and the LET joint c.
(2) Deformed configuration of rotation, d.

		KM
M-25	CORE Bearing	RTL

(1)

This mechanism is a compliant rolling-contact element (CORE). The CORE bearing is designed by combining the basic CORE elements in a way that allows a rotational motion. The CORE bearing imitates a planetary gear system with three planets, a sun and a ring [18].

(1) Rigid body a, b and c are the ring, sun and planets, respectively. Allowing rotation about the d-axis.

M-26   **Long-Stroke Flexure Pivot**

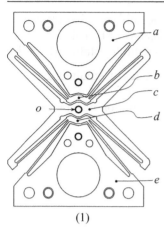

(1)

(2)

This flexure pivot has a large angular stroke (typically ±15°) and very small center shift (typically 1 micrometer). This planar and monolithic pivot is composed of 4 remote-center-compliance spring pivots placed in series. This arrangement leads to a total angular stroke that is 4 times that of a single pivot. Moreover, each pair of pivots benefits from a center-shift compensation effect. As a result, the complete structure has a very small residual center-shift.

(1) The fixed base e is connected to a first intermediate block d via a pair of plate-springs that form a pivot about the axis o. A second pair of plate-springs links d to the second intermediate block c, producing a pivot about the same axis o and compensating most of the parasitic shift of the first stage. A third pair of plate-springs lead to b and a fourth pair to the output block a on which the payload is to be attached.
(2) Photo of such a pivot designed for aerospace applications.

See also

Name	Reference Index	Categorization Index
CORE	EM-10	FE/KM FRH/RTL
Isolation-based HCCM	EM-20	FE/KM FRH/RTL
Inversion-based HCCM	EM-21	FE/KM FRH/RTL
Inversion-based HCCM	EM-22	FE/KM FRH/RTL
HCCM	EM-23	FE/KM FRH/RTL

Precision
See also

Name	Reference Index	Categorization Index
Ortho Skew Double Rotary	EM-42	FE/KM FU/RTO
Tripod	EM-43	FE/KM FU/RTO

12.2.3 Translation—Rotation

M-27   **Mechanically Actuated Trigger Switch**   KM
                                                                                    TR

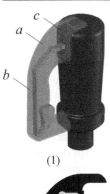

(1)

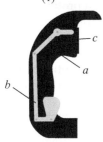

(2)

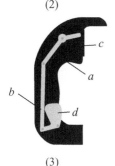

(3)

This is a mechanically actuated compliant trigger that can be integrated into the removable handle of the RotoZip spiral saw, a multifunctional tool that is capable of performing a wide variety of construction related tasks through the use of available attachments [19].

(1) The figure shows a compliant mechanism b housed within the removable handle a of the RotoZip spiral saw. The removable handle covers the bump switch c, providing tool lock-out by isolating the bump switch from the user.

(2) The figure shows the bump switch c in the off position, wherein the tool would not be powered.

(3) The figure shows the bump switch c in the on position. When the trigger d is pressed, the coupler segment of the compliant mechanism b, while remaining in tension, activates the bump switch to power the tool.

	Compliant Double-Arm	KM
M-28	Suspension Mechanism	TR

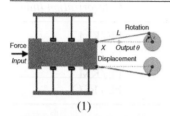

(1)

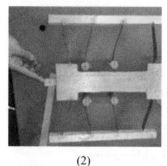

(2)

Double Parallel Arm Suspension

(3)

A double-arm compliant mechanism consists of eight large deflecting flexible arms and a rigid coupler, and can be actuated by an electrostatic comb drive. The aim of the mechanism is to transform linear motion into rotational motion. The linear motion is achieved by the parallel double arm mechanism and the rotational part is achieved by the crank part of the slider-crank mechanism. Trajectory control is achieved by state feedback since the compliant double-arm mechanism has a nonlinear stiffness. A PD (Proportional Derivative) controller scheme is used and the coefficients of the controller are found by using the desired specifications expected from the output of the system [20].

(1) Slider-crank mechanism driven by a compliant double arm mechanism.
(2) Deformed macrocompliant double parallel arm mechanism prototype made for visualization purposes.
(3) The translational motion of the shuttle suspended above the ground can be converted to a rotational motion with the addition of a coupler and a crank pair.

Precision

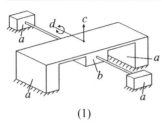

(1)

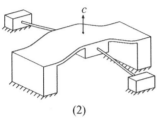

(2)

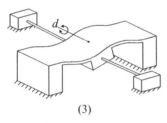

(3)

This mechanism is suited for precision applications where translational and rotational degrees of freedom are required. The translational and rotational degrees of freedom are orthogonal. The translational degree of freedom is orthogonal to the plane of the ground. The rotational degree of freedom is parallel to the plane of the ground [13].

(1) Rigid bodies a are fixed. Rigid body b translates in the c-direction and rotates about the d-axis.
(2) Deformed configuration of translation in the c-direction.
(3) Deformed configuration of rotation about the d-axis.

M-30	Crossed Constraint Translator and Rotator	KM TRP

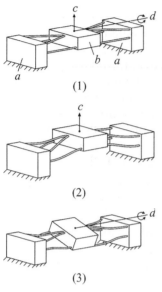

(1)

(2)

(3)

This mechanism is suited for precision applications where both a translational and rotational degree of freedom is required. The translational degree of freedom is orthogonal to the axis of the rotational degree of freedom. The translational degree of freedom is orthogonal to the plane of the ground and the rotational degree of freedom axis is parallel to the plane of the ground [13].

(1) Rigid bodies a are fixed. Rigid body b translates in the c-direction and rotates about the d-axis.
(2) Deformed configuration of translation in the c-direction.
(3) Deformed configuration of rotation about the d-axis.

M-31	Precision Octa Parallel Symmetric Constraint	KM TRP

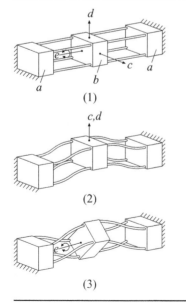

(1)

(2)

(3)

This mechanism is suited for precision applications where two orthogonal translational degrees of freedom and one rotational degree of freedom are required. All of these degrees of freedom are orthogonal. The two translations are parallel to the planes of the grounds and the rotation is perpendicular to the planes of the grounds [13].

(1) Rigid bodies a are fixed. Rigid body b translates in the c- and d-direction and rotates about the e-axis.
(2) Deformed configuration of translation in the c- or d-direction.
(3) Deformed configuration of rotation about the e-axis.

M-32   **Octa Constraint Rotary**

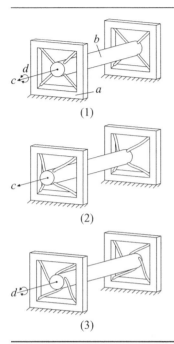

(1)

(2)

(3)

This mechanism is suited for precision applications that require a translational degree of freedom and one rotational degree of freedom [13].

(1) Rigid bodies a are fixed. Rigid body b translates in the c-direction and rotates about the d-axis.
(2) Deformed configuration of translation in the c-direction.
(3) Deformed configuration of rotation about the d-axis.

M-33   **Parallel Blade Constraint**

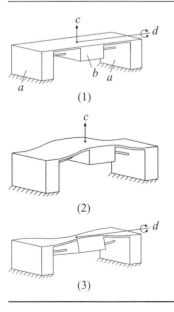

(1)

(2)

(3)

This mechanism is suited for precision applications where rotational and translation degree of freedom are required. The rotational and translational degree of freedoms are orthogonal. The translational degree of freedom is perpendicular to the plane of the ground and the rotational degree of freedom is parallel to the plane of the ground [13].

(1) Rigid bodies a are grounded. Rigid body b translates in the c-direction and rotates about the d-axis.
(2) Deformed configuration of translation in the c-direction.
(3) Deformed configuration of rotation about the d-axis.

Large Motion Path

M-34	Quadra Parallel Constraint	KM TRL

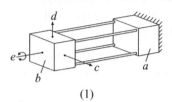

(1)

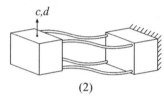

(2)

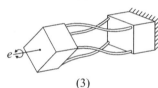

(3)

This mechanism is suited for precision applications where two orthogonal translational degrees of freedom and one rotational degree of freedom are required. All of these degrees of freedom are orthogonal. The two translations are parallel to the plane of the ground and the rotation is perpendicular to the plane of the ground. The rotation will cause the rigid body to retract toward the ground with an undesired translation if the rotation is not small enough [13].

(1) Rigid body a is grounded. Rigid body b may translate in the c- and d-directions and rotate about the e-axis.
(2) Deformed configuration of translation in c- or d-direction.
(3) Deformed configuration of rotation about the e-axis.

12.2.4 Parallel Motion

M-35	4-Bar Parallel Guider	KM PM

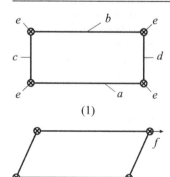

(1)

(2)

A 4-bar, parallel-guiding mechanism is a mechanism whose two opposing links remain parallel throughout the mechanism's motion. This design can have multiple configurations based upon its synthesis [21, 22, 23].

(1) Rigid body a is fixed. Points b, c, d, and e provide a pivot rotation by either a flexible or rigid element.
(2) Deformed configuration in the f-direction.

M-36   **Parallel Guiding Optic**

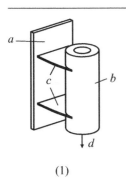

(1)

This mechanism utilizes parallel-guiding mechanisms for an optic to focus, using a fully compliant mechanism. Doing this helps with backlash and also allows the optic to stay perpendicular to the piece the optic is focused on [21, 24].

(1) Rigid body a is fixed. Rigid body b, the optic, translates in the d-direction by flexible segments c.

M-37   **Press**

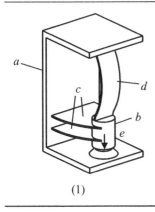

(1)

This mechanism utilizes a parallel-guiding mechanism and a buckled flexible member. When the flexible member is straightened it translates the press, guided by the parallel-guided component [21, 24].

(1) Rigid body a is fixed. Rigid body b translates in the e-direction by flexible segments c when a force/displacement is inputed on the flexible segment d.

Precision

M-38	**Parallel Guiding**	KM PM

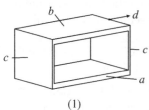

(1)

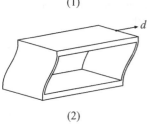

(2)

This mechanism achieves this motion by two fixed-guided beams [22, 24].

(1) Rigid body *a* is fixed. Rigid body *b* translates in the *d*-direction by the flexible fixed-guided beams *c*.
(2) Deformed configuration of translation in the *d*-direction.

Large Motion Path

M-39	**Parallel Guided**	KM/KN PML/ES

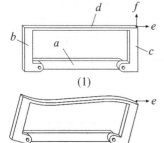

(1)

(2)

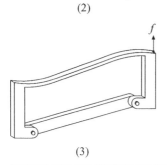

(3)

This mechanism's links remain parallel throughout the mechanism's motion and is capable of large deflections with energy storage [25].

(1) Rigid body *a* is a rigid link. Rigid bodies *b* and *c* are rigid segments. Segment *d* is a fixed-guided beam.
(2) When rigid body *a* is fixed, the mechanism deforms in the *e*-direction.
(3) When rigid body *b* is fixed, the mechanism deforms in the *f*-direction.

M-40   **Parallel-Guided LEM**

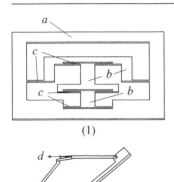

(1)

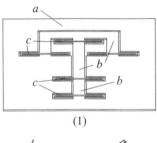

(2)

This is a lamina emergent parallel-guiding mechanism. It achieves its motion through torsion elements and LET joints. This mechanism can be fabricated in a single plane [26].

(1) Rigid body *a* is fixed. Rigid bodies *b* are rigid-link segments. Flexible segments *c* provide the rotational motion.
(2) Deformed configuration in the *d*-direction.

M-41   **Parallel-Guided LEM**

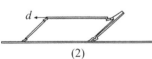

(1)

(2)

This is a lamina emergent parallel-guiding mechanism. It achieves its motion through torsion elements and LET joints. This mechanism can be fabricated in a single plane [26].

(1) Rigid body *a* is fixed. Rigid bodies *b* are rigid-link segments. Flexible segments *c* provide the rotational motion.
(2) Deformed configuration in the *d*-direction.

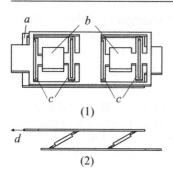

(1)

(2)

This is a multi-layer, lamina emergent, parallel-guiding mechanism. It achieves its motion through torsion elements and LET joints. This mechanism can be fabricated in a single plane [26].

(1) Rigid body *a* is fixed. Rigid bodies *b* are rigid-link segments. Flexible segments *c* provide the rotational motion.
(2) Deformed configuration in the *d*-direction.

12.2.5 Straight Line

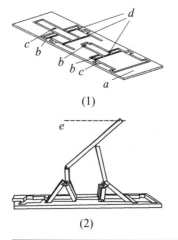

(1)

(2)

This is a fully compliant lamina emergent Hoeken mechanism that was designed using the compliant ortho-planar metamorphic mechanism (COPMM) technique. A Hoeken mechanism produces a straight line through part of its motion. This mechanism can be fabricated in a single plane [27, 28].

(1) Rigid body *a* is fixed. Segment *b* is inserted into segment *c* during assembly. Segments *d* allows flexibility.
(2) Assembled configuration of mechanism. The end point traces through a near straight line, *e*.

KM
SL

M-44   **Monolithic Straight-Line Tracer**

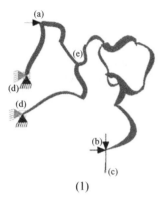

(1)

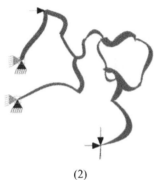

(2)

(3)

A monolithic mechanism that traces a straight-line approximately [29].

(1) A single-piece approximate straight line tracer obtained via topology, shape and size optimization. *(a)* horizontal rightward input force. *(b)* tracing point. *(c)* the straight-line path. *(d)* fixed supports. *(e)* continuum composed of initially curved deforming members.
(2) An intermediate configuration of the straight-line tracer.
(3) Final deformed configuration of the line tracer.

12.2.6 Unique Motion Path

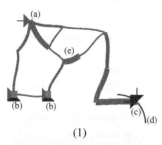

(1)

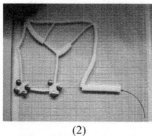

(2)

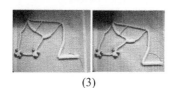

(3)

The bottom right port of the monolithic (fully compliant or single-piece continuum) mechanism traverses a circular arc if the top left port is actuated by a horizontal force [29].

(1) Design of a single-piece arc tracer obtained via topology, shape and size optimization in simulation. (a) horizontal input force. (b) fixed supports. (c) tracing point. (d) output path. (e) continuum composed of initially curved deformable members.

(2) Prototype of the arc tracer fabricated using ABS thermoplastic.

(3) Various displaced configurations of the arc tracer.

M-46   **Partially Compliant Tick Path Tracer**

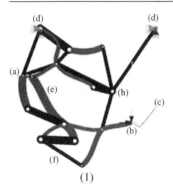

(1)

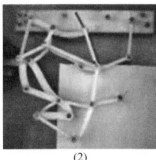

(2)

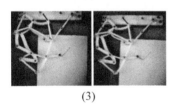

(3)

A point on this partially compliant mechanism traces a tick path when another point on this linkage is actuated by a monotonically increasing horizontal force. A compact rigid-link mechanism with similar topological simplicity is difficult to design [30].

(1) Design of a tick path tracer via link-type, topology, shape, and size optimization obtained in simulation. (*a*) port of actuation wherein a leftward horizontal force acts. (*b*) the point on the mechanism tracing the tick path. (*c*) the tick path; desired and actual paths are superposed. (*d*) fixed supports. (*e*) deformable members modeled as initially curved large deformation frame finite elements. (*f*) rigid links with hinges at the two ends. (*h*) hinges depicted by white circles.

(2) Prototype of the tick path tracer fabricated using ABS thermoplastic. Both deformable and rigid links are manufactured using the same material.

(3) An intermediate and final displaced configurations of the tick path tracer.

M-47   **Partially Compliant Hat Path Tracer**

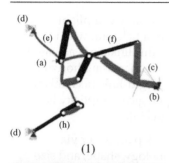

(1)

(2)

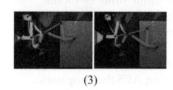

(3)

A point on this partially compliant
mechanism traces a desired hat path if
another point is pushed horizontally to
the right [30].

(1) Mechanism obtained via link-type,
topology, shape, and size
optimization. (*a*) port of actuation. A
horizontal force directed towards the
right moves the mechanism. (*b*) the
output point tracing the tick path.
(*c*) the tick path; desired and actual
paths are overlapped. (*d*) fixed
supports. (*e*) members undergoing
large deformation. (*f*) rigid links. (*h*)
hinges shown as white circles.
(2) ABS prototype of the hat-path tracer
with both deformable and rigid links
manufactured using the same material.
(3) Intermediate and final displaced
configurations of the hat-path tracer.

M-48	Monolithic Contact-Aided Compliant Ramp-Path Tracer #1	KM UP

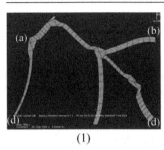

(1)

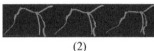

(2)

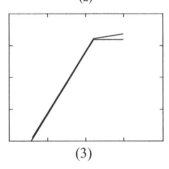

(3)

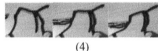

(4)

This ramp-tracing monolithic mechanism does not employ individual deformation/rotation contributions from its flexible and rigid members. Instead, it uses only intermittent contact between its deforming members to generate a desired kink on the path. The mechanism was designed through a structural topology design method that not only determined the constituents, but also resolved which (two or more) members interact with each other. It also determined when and for what duration the members are in contact. For the same application, the topology is deemed simpler compared to a partially compliant counterpart [31].

(1) The ramp-path tracing mechanism.
   (a) port of rightward horizontal input.
   (b) output port tracing the ramp path.
   (d) fixed supports.
(2) Three intermediate deformed configurations.
(3) The path traced by point (b) in Fig. (1). Desired and delivered ramp paths are overlapped.
(4) The prototype constituted of rubber shown in three deformed configurations.

M-49	**Monolithic Contact-Aided Compliant Ramp-Path Tracer #2**	KM UP

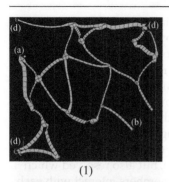

(1)

(2)

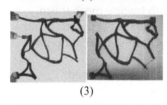

(3)

The second ramp-tracing monolithic contact-aided compliant mechanism [31].

(1) Design obtained using structural topology optimization. *(a)* port of rightward horizontal actuation. *(b)* output port. *(d)* fixed supports. Compared to M-48, this mechanism is designed using a finer supergrid structure.
(2) Intermediate deformed profiles of the ramp tracer.
(3) A rubber prototype of the ramp-path tracer shown in a deformed configuration. The prototype is shown in its undeformed profile. The ramp-path traced is also illustrated.

| M-50 | Compliant Non-smooth Path Generator
with Smooth Input | KM
UP |

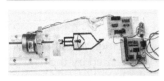

(1)

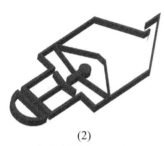

(2)

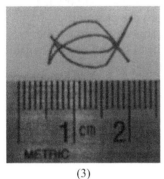

(3)

The first figure shows a working prototype of a contact-aided compliant mechanism. Its input point is attached to a screw driven by a stepper motor. It has two output points where pointed needles are attached. They trace a pair of enclosing paths with cusps and non-smooth points when the input point is reciprocated smoothly along a straight line. Intermittent contact occurring at points marked with arrows gives this mechanism this unusual behavior. It can be used to tease a cell out of tissue by repeatedly tracing the paths with a sharp needle [32].

(1) The prototype of a non-smooth path-generator.
(2) Solid model of the compliant mechanism.
(3) A pair of nonsmooth paths enclosing an area.

M-51	**Fully Compliant Five Bar Mechanism Design and Control for Trajectory Following**	**KM UP**

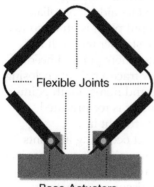

Flexible Joints

Base Actuators

(1)

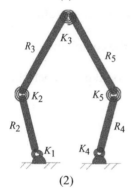

(2)

A fully compliant five-bar mechanism consists of five rigid segments connected by flexible joints. It is excited by the applied torques at the base links. The mechanism is synthesized to achieve desired trajectories.

(1) Compliant five-bar mechanism with rigid segments and flexible joints.
(2) SPRBM of the compliant five-bar mechanism. The large deflecting elastic joints are represented by the torsional springs.
(3) The purpose of controlling the compliant mechanism is to track a path within a work area of the mechanism. This figure shows the trajectory control output of a signature and the reference signature.

Trajectory Control of Signature

— reference
--- Out

Vertical displacement [m]

0.42
0.41
0.4
0.39
0.38

0.26 0.27 0.28 0.29 0.3 0.31 0.32 0.33 0.34 0.35
Horizontal displacement [m]

(3)

12.2.7 Stroke Amplification

	Slaving Mechanism for Compound	KM
M-52	Flexure Pivots	SA

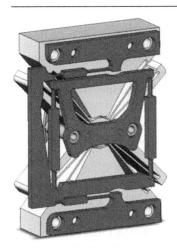

(1)

(2)

A common solution to increase the stroke of flexure pivots is to stack two identical pivots with concomitant rotation axes in series. This approach leads to an internal degree of freedom that is undesired in some situations (e.g. fast motions and/or high external radial loads). A slaving mechanism might be used to suppress this undesired degree of freedom. This mechanism kinematically links the base, the output and the intermediate block with a 1:2 motion law. This law is a result of the symmetry of this mechanism. Note: This slaving mechanism fulfills the same role as the slaving lever that is commonly used with the compound parallel spring stage.

(1) Slaving mechanism (thin plate) mounted on top of a butterfly flexure pivot.
(2) Slaving mechanism integrated monolithically in a compound remote-center compliance pivot

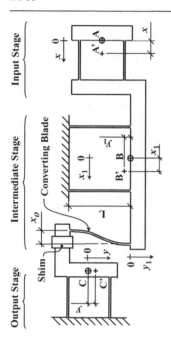

(1)

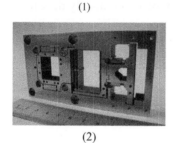

(2)

This compliant mechanism converts an input motion into a smaller perpendicular output motion with a large and constant reduction ratio (typically 1:100 to 1:1000). Working principle: an external actuator drives point A rectilinearly to A'. This motion is transmitted to the Intermediate Stage that is guided by a classical parallel-spring-stage (with blade length L): point B moves to B'. Due to the shortening of the blade projection, the motion of this stage is a well-known parabolic translation. A third blade of same length L (called "Converting Blade"), that has an offset deformation x_0, links the Intermediate Stage to the Output Stage. The Output Stage is guided vertically by a classical parallel-spring-stage. The motion x_1 causes the Converting Blade to shorten, following the same parabolic law as the two blades of the Intermediate Stage, but with an offset x_0. The resulting motion y of the Output Stage (motion from C to C') is equal to the differential shortening of the blades (subtraction of two parabolas with an offset). As a result, the displacement of the Output Stage is simply proportional to the displacement x of the actuator, with reduction ratio i that is constant over the whole displacement range and is inversely proportional to the offset x_0. $i = \frac{x}{y} = \frac{5L}{6x_o}$.

(1) Working principle of the Nanoconverter
(2) Photograph of a monolithically machined Nanoconverter used for a Differential-Phase-Contrast Interferometer on a synchrotron beamline.

		KM/KN
M-54	**Pantograph (LEM)**	SA/FA

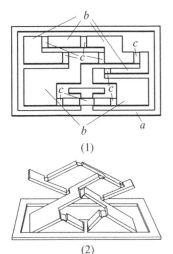

(1)

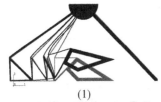

(2)

A pantograph mechanism is a multi-degree of freedom device used for scaling force or motion. This mechanism is designed to be lamina emergent [33].

(1) Rigid body *a* is fixed. Segments *b* are significantly rigid and segments *c* allow flexibility.
(2) Deformed configuration.

	Displacement-Amplifying	KM
M-55	**Compliant Mechanism**	SA

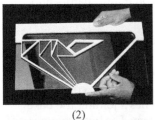

(1)

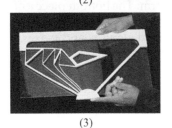

(2)

(3)

The first figure shows a displacement-amplifying compliant mechanism in its original and deformed configurations. The next two figures show the prototype in two configurations [34].

(1) Simulated displacement-amplifying compliant mechanism.
(2) Prototype when the applied force is small.
(3) Prototype when the applied force is large.

12.2.8 Spatial Positioning

		KM/KN
M-56	**Multiple Stage Platform**	SP/ES

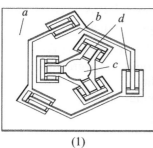

(1)

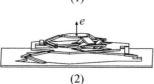

(2)

This mechanism is similar to ortho-planar springs, but it uses a multi-stage platform to raise its platform [33].

(1) Rigid body *a* is fixed. Rigid bodies *b* and *c* are platforms. Segments *d* are the flexible segments allowing platform *c* to translate in the *e*-direction.
(2) Deformed configuration after translation in the *e*-direction.

		KM
M-57	**Compliant Parallel Platform**	SP

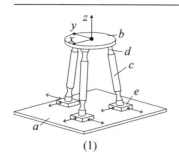

(1)

This is a monolithic fully compliant platform mechanism that is actuated by six linear actuators on the base. It can be applied to precision manipulators and positioners [35].

(1) The base link *a* is fixed. The moving platform *b* is connected by three compliant limbs. Each limb is formed by a rigid link *c* and two compliant spherical notches *d* joining the base and the platform. The bottom piece *e* of each limb is actuated with two linear actuators aligned along *x*- and *y*-axes.

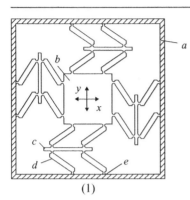

(1)

This monolithic fully compliant platform mechanism is designed to position the center stage along x- and y-axes [35, 36].

(1) The outer frame a is fixed. The moving stage b is connected to a via four identical double parallelograms c that are in symmetrical layout. Formed by two parallelograms connected in series, each double parallelogram constrains the stage to translate along x- and y-axes. All parallelograms are formed with two rigid links d and four flexure joints e.

Precision

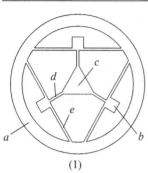

(1)

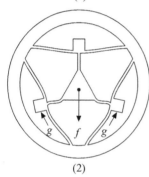

(2)

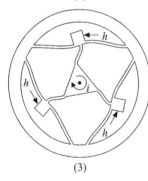

(3)

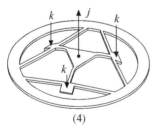

(4)

The HexFlex™ is a single-layer, multi-axis spatial positioning control mechanism, which can be used for both macro- and microapplications that require precision positioning [37].

(1) Rigid body *a* is fixed. Rigid bodies *b* are the actuator action tabs. Rigid body *c* is the motion stage. Flexible elements, *d* and *e*, allow infinitesimal motions.
(2) Deformed configuration by planar displacement of the actuator tabs in the *g*-direction, which causes the motion stage to displace in the *f*-direction.
(3) Deformed configuration by planar displacement of the actuator tabs in the *h*-direction, which causes the motion stage to rotate about the *i*-axis.
(4) Deformed configuration by orthogonal displacement of actuator tabs in the *k*-direction, which causes the motion stage to translate in the *j*-direction.

M-60	**Zero-Stiffness 6-DoF Compliant Precision Stage**	KM/KN SPP/SBB

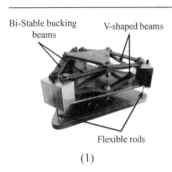

Bi-Stable bucking beams   V-shaped beams

Flexible rods

(1)

This mechanism is a six-degrees of freedom zero-stiffness mechanism with a near-zero actuation force in a finite range of motion, able to balance a constant force. The three out-of-the-horizontal-plane motions are performed by cooperative action of bistable beams and v-shaped beams; the three in-the-horizontal-plane motions are performed by three flexible rods, with zero stiffness for the three in-plane motions when loaded to the buckling load [38].

(1) Prototype of the zero-stiffness 6-DoF precision stage.

12.2.9 Metamorphic

M-61	**Lamina Emergent four-Bar**	KM MM

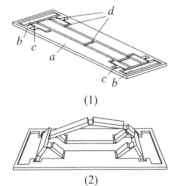

(1)

(2)

This is a lamina emergent four-bar mechanism that was designed using the compliant ortho-planar metamorphic mechanism (COPMM) technique, allowing the mechanism to be raised from the initial plane of fabrication by using a system of redundant link structures [27].

(1) Rigid body a is fixed. Segments b are inserted into segments c during assembly. Segments d allow flexibility from the manufactured state to the configured state.
(2) Assembled configuration of mechanism.

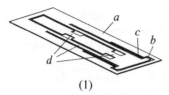

(1)

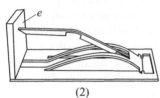

(2)

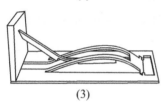

(3)

This is a bistable locking compliant ortho-planar metamorphic mechanism (COPMM) [27].

(1) Rigid body *a* is fixed. Segment *b* is inserted into segment *c* during assembly. Segments *d* allow flexibility.
(2) Assembled and stable configuration of mechanism. Rigid body *e* is attached to a mechanism during assembly.
(3) Deformed and stable configuration of mechanism.

M-63	COPMM Bistable Switch	KM/KN MM/SBB

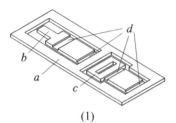

(1)

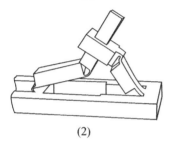

(2)

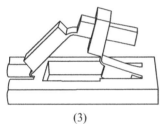

(3)

This is a compliant ortho-planar metamorphic mechanism (COPMM) that is a bistable switch. It is based on a fully compliant switch, and was redesigned by using the COPMM technique. This switch can be manufactured in a single plane and is assembled out of the plane for operation [27].

(1) Rigid body a is fixed. Segment b is inserted into segment c during assembly. Segments d allow flexibility.
(2) Assembled and stable configuration of mechanism.
(3) Deformed and stable configuration of mechanism.

M-64	**Bistable COPMM**	KM/KN MM/SBB

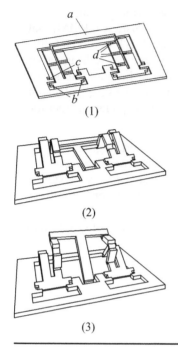

(1)

(2)

(3)

This is a fully compliant ortho-planar metamorphic mechanism (COPMM) that is bistable. It is based upon a closed-loop 6-bar to a bistable non-grashoffian 4-bar [27].

(1) Rigid body *a* is fixed. Segment *b* is inserted into segment *c* during assembly. Segments *d* allow flexibility.
(2) Assembled and stable configuration of mechanism.
(3) Deformed and stable configuration of mechanism.

See also

Name	Reference Index	Categorization Index
Hoeken (LEM)	M-43	KM SL/MM

12.2.10 Ratchet

M-65   **Overrunning Ratchet Clutch**

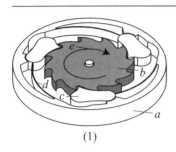

(1)

This mechanism is an overrunning ratchet and pawl clutch with centrifugal throw-out. An important factor in the design is the use of passive joint elements that allow rotation of the pawls [39].

(1) Rigid body a is fixed. Rigid-body b rotates in the e-direction. Rigid-bodies c, the pawls, prevents rotation in the opposite direction. The pawls are able to deflect by using the flexible segments, d, and resist motion by using a passive element. The extra mass on the pawls, c, allows the centrifugal throw-out.

M-66   **X Bob Ratchet**

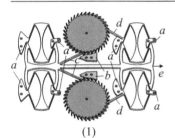

(1)

This mechanism integrates the X Bob into a ratcheting system. This mechanism is designed so that the wheel advances one and only one tooth per actuation, in planar arrangement, and can be fabricated in the microscale [3].

(1) Rigid bodies a and b are fixed. Rigid body b only allows the wheel to advance one tooth per actuation. Flexible segments c actuate the wheel and flexible segments d prevent motion in the opposite direction.

KM
RC

M-67   **X Ratchet**

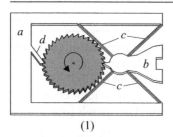

(1)

This mechanism uses cross-flexures that allow a rotational degree of freedom, which is used to actuate the wheel.

(1) Rigid body *a* is fixed. Rigid body *b* actuates the wheel by the flexible segments *c*. Flexible segment *d* prevents motion in the opposite direction.

KM
RC

M-68   **CHEQR**

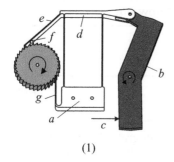

(1)

This mechanism is a high-precision e-quintet ratcheting (CHEQR) mechanism. This mechanism is designed so that the wheel advances one and only one tooth per actuation, in a planar arrangement, and can be fabricated in the microscale [40].

(1) Rigid body *a* is fixed. Rigid link *b* actuates the parallel-guiding component *d*, by an input *c*. Flexible segment *e* actuates the wheel and rigid body *f* only allows one tooth advancement per actuation. Flexible segment *g* prevents motion in the opposite direction.

M-69   **Triggering Ratchet**

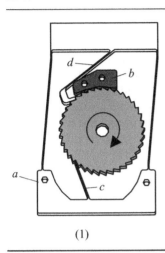

(1)

This ratcheting mechanism is designed so that the wheel advances one and only one tooth per actuation, in planar arrangement, and can be fabricated in the microscale [40].

(1) Rigid bodies *a* and *b* are fixed. Flexible segment *d* actuates the wheel and rigid body *b* only allows one tooth advancement per actuation. Flexible segment *c* prevents motion in the opposite direction.

M-70   **RaPR**

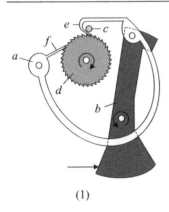

(1)

This mechanism is a ratchet and pawl ring (RaPR) mechanism. This mechanism is designed so that the wheel advances one and only one tooth per actuation, in planar arrangement, and can be fabricated in the microscale [41].

(1) Rigid body *a* is fixed. Rigid link *b* actuates the pawl ring component *d*. Flexible segment *e* actuates the wheel and pin *c* only allows one tooth advancement per actuation. Flexible segment *f* prevents motion in the opposite direction.

(1)

(2)

(3)

Pedaling in one direction takes the bicycle forward; the large sprocket wheel freewheels when pedaled in the opposite direction. This is conventionally achieved with many parts including a ratchet. The figures show an alternate arrangement in which two identical parts with two compliant cantilevered pawls riding over the internal sprocket wheel [42].

(1) Sprocket wheel and a pair of parts with two cantilevered pawls.
(2) One part fitted into the sprocket wheel.
(3) Both parts fitted into the sprocket wheel.

12.2.11 Latch

<table>
<tr><td>**M-72**</td><td style="text-align:center">Latching</td><td style="text-align:right">KM
LC</td></tr>
</table>

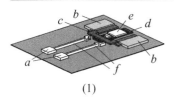

(1)

This mechanism is a two-position microlatching mechanism (MLM) requiring a single actuator. The MLM maintains its second position from a mechanical interference between a slider and flexible beams. The lock slider receives the input and latches. The wedge slider allows the release of the lock slider [43].

(1) Rigid bodies *a* and *b* are fixed. The wedge slider *c* slides through the anchor *b*, and controls the lock slider *d* by stops *e*. The lock slider is inserted into the flexible members *f* and is released by the wedge slider.

M-73	Fluid Level Indicator Locking Mechanism	KM LC

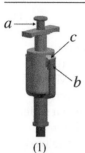

(1)

(2)

(3)

(4)

This is a locking mechanism for a fluid level indicator (FLI), for Orscheln Products, LLC, capable of enduring exposure to extreme temperatures, motor fluids, and ultraviolet rays. The design is to be one with a simple, intuitive means for locking and removal [44].

(1) Cylindrical T-button a is subjected to a small vertical displacement. The compliant arms b deform outward and then lock around the face c.
(2) Cross-sectional view of the T-button FLI mechanism.
(3) T-button in unlocked position.
(4) T-button in locked position. The parts align themselves rotationally as they mate using a cam-follower arrangement.

12.2.12 Others

		KM
M-74	Self-Adaptive Finger	KMO

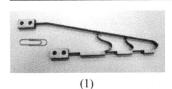

(1)

(2)

(3)

A fully compliant three-phalange finger with one actuation site (underactuation) adapts to various objects without any sensor. The compliant network distributes the actuation force over the phalanges [45].

(1) Side view of the finger. Three flexures can be seen in the bottom chain. The top chain with the S-curved segments distributes the operating force.
(2) Partially compliant gripper consisting of two compliant fingers and a differential linkage.
(3) Closing sequence shows adaptation to external influences.

	Compliant, Circumferentially Actuated,	KM
M-75	Radially Deployable Mechanism	KMO

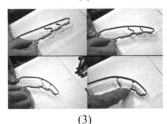

(1)

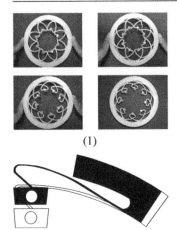

(2)

The set of four figures show a two-layered but one-piece compliant mechanism that can move in and out radially when its two rings in two different layers are rotated relative to each other. Its compliant mechanism is shown in the second figure where circumferential motion is amplified in the radial direction. It is useful in holding circular or any regular-polygon shaped objects with uniform force [46].

(1) Circumferentially actuated, compliant, radially deployable mechanism in its four configurations.
(2) Compliant element underlying the mechanism.

		KM
M-76	**Compliant Cycle-Doubler**	KMO

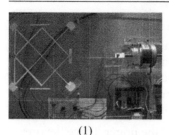

(1)

(2)

The figure shows the working prototype of a one-piece compliant mechanism that can double the cycles of a reciprocating translational input. Its mechanism shown in the second figure has a contact-aided compliant mechanism. It needs two rigid surfaces (CS) where contact points (CN) touch and change the direction of the output path of point OP while the input point IP continues to move in the same direction in one half of its reciprocating motion [47].

(1) The prototype of a compliant cycle-doubler.
(2) Schematic of the cycle-doubling compliant mechanism.

		KM
M-77	**Cam Flexure**	KMO

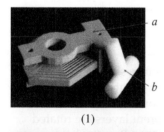

(1)

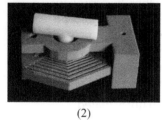

(2)

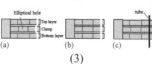

(3)

The two figures show a cam-flexure mechanism that can hold a small capillary tube of circular cross-section used in wire-bonding machines. The T-shaped tool is to be inserted into an elliptical hold and turned to align the holes [see Figure (3)]. When released, the tube is held tightly [48].

(1) The cam flexure a and the tool b.
(2) The cam flexure with the tool inserted.
(3) Schematic of the cam flexure.

12.3 Kinetics

12.3.1 Energy Storage

		KN
M-78	**Leaf Spring**	ES

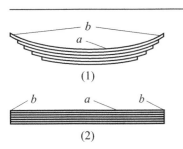

(1)

(2)

Leaf springs can be designed to provide a desired motion function, however, their primary function is that of a spring. These types of springs use a stacking approach to reduce the space and weight of the mechanism while maintaining functionality [49].

(1) A typical leaf-spring configuration where the length of the flexures, *a*, vary. *b* is where the springs are mounted.
(2) Another configuration where the length of the flexures, *a*, are the same. *b* is where the springs are mounted.

		KN
M-79	**Ortho–Planar Spring**	ES

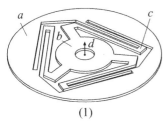

(1)

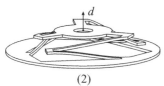

(2)

This mechanism is an ortho-planar spring that operates by raising and lowering its platform in relation to the base. The benefit of this mechanism is it achieves this motion without rotation, which eliminates problems of rotational sliding against adjoining surfaces and has less sensitive variation in assemblies [50, 51].

(1) Rigid body *a* is fixed. Rigid body *b* is the platform, which translates in the *d*-direction through the flexible switch backs, *c*.
(2) Deformed configuration of translation in the *d*-direction.

KN

M-80   **Rhombus Spring**   ES

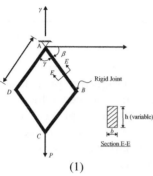

(1)

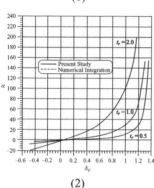

(2)

A nonlinear spring action could be obtained from a compliant rhombus frame [52].

(1) Four compliant links are fixed to each other to form the rhombus frame with a specific apex angle γ. The links could be tapered by linearly varying the cross-sectional height h by specifying a taper ratio t_r between the values of h at the two ends of the link. The frame is pinned at A and a force is applied at C as shown.

(2) The nonlinear deflection behavior of the spring for an apex angle $\gamma = 150°$ and different taper ratios $t_r = 0.5, 1.0$ and 2.0 is shown. The applied force is defined by a nondimensional parameter α combining the applied load P with the links flexural rigidity and length. It is observed that a soft spring action is obtained at low loads and a sudden hardening action occurs at certain deflection values.

KN

M-81   **Monolithic Stapler**   ES

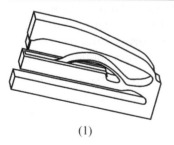

(1)

The figure shows a one-piece compliant stapler. The staple-loading slot, the compliant segment that hold the staple-stack tight, two flexure joints, and the plunger that pushes the staples are all integrated into a one injection-moldable part. The die to fold a staple around a stack of papers is also incorporated on the rigid beam on the bottom [53].

(1) Schematic.

See also

Name	Reference Index	Categorization Index
Parallel Guided	M-39	KM/KN PML/ES
Multiple Stage Platform	M-56	KM/KN SP/ES

Clamp

M-82	**Gripper Hook**	KN ESC

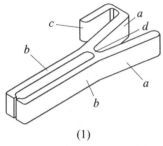

(1)

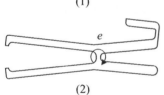

(2)

This mechanism uses a small length flexure as the pivot and the energy-storage device [54].

(1) When rigid bodies *a* come together, rigid bodies *b* separate by the flexible segment *d*. Rigid body *c* is a hooking device.

(2) Deformed configuration.

M-83	**Gripper**	KN ESC

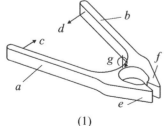

(1)

This mechanism uses a small length flexure as the pivot and the energy storage device.

(1) Rigid segment *a* and *b* translate in the *c*- and *d*-direction, respectively, causing rigid segments *e* and *f* to separate. Flexible segment *g* allows the deformation and stores the energy.

M-84   **Clamp**

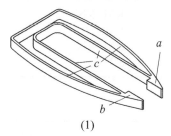

(1)

This mechanism uses flexures for deformation and energy storage device [54].

(1) Rigid segment *a* and *b* are the clamping surfaces. Flexible segments *c* allows the deformation and stores energy to provide the clamping force.

M-85   **Clamp 3D**

(1)

The figure shows a 3D compliant gripper that has three halves of a 2D gripper arranged at 120° apart. When the central portion is pressed down, the three jaws of the gripper come together to hold an object.

(1) 3D compliant gripper.

(1)

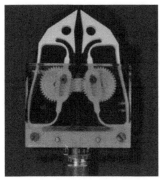

(2)

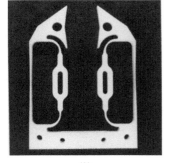

(3)

The compliant gripper mechanism provides a near-parallel grasp. With a modest re-design, it could be enhanced to provide a gripping force that would have a nearly constant value [55].

(1) This figure shows the compliant gripper mechanism in its open position.
(2) This figure shows the compliant gripper mechanism in its near-closed position.
(3) This figure shows the enabling, fully compliant (one-piece) grasping fingers mechanism fabricated from a high-stiffness and resilient material such as Nylon and Delrin®.

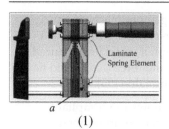

Laminate
Spring Element

a

(1)

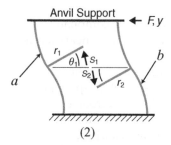

Anvil Support   ← *F, y*

r_1

θ_1 ↑s_1 *b*

s_2↓

a r_2

(2)

This force-sensing compliant clamp (Sense Clamp) is a clamping device with integrated compliant laminae to measure the clamping pressure applied. A needle attached to the outer two laminae is calibrated to read the load exerted. Simple construction, inexpensive and good feel of this product adds to its functionality. The ability of the clamp to sense the applied force allows a user to consistently and evenly clamp together a workpiece, especially if more than one clamp is required for an application [56].

(1) This figure shows a 3D rendering of the sense clamp, with *a* pointing to the calibrated sensing needle.
(2) *a* and *b* are the two outer laminae. The two lever arms r_1 and r_2 rotate in opposing directions creating the input displacement necessary to actuate and calibrate the needle mechanism.
(3) This figure shows a working prototype of the force-sensing compliant clamp.

Sen e Clamp

(3)

M-88	Partially Compliant Displacement Delimited Gripper Mechanisms	KN ESC

(1)

(2)

(3)

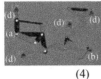

(4)

Partially compliant displacement delimited gripper mechanisms can be used to grip very soft work-pieces (e.g., biological cells) to prevent damage due to excessive force or pinching [57].

(1)–(4) Different top-symmetric designs and their full prototypes. (a) horizontal leftward actuation. (b) the gripper ports traversing the "J" paths. (d) fixed supports. The workpiece is to be gripped at the bottom, horizontal portion of the "J" path. Because there is no significant vertical deflection of the output port, the reaction force on the workpiece is negligible. Of these four designs, the design in (2) is the best.

M-89	Compliant Gripping Device	KN ESC

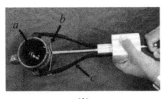

(1)

This compliant mechanism features an early, basic, and passive gripping device that was designed for use by a person with a missing hand. A simple and inexpensive device, sometimes described as a poor man's hand, it is operated by a wire rope reaching back and tied to a braced torso. It can add much needed mobility at low expense to improve the quality of life [58].

(1) The figure shows the compliant gripping mechanism with jaws *a* lined with high friction material, compliant members *b* made from thermoplastic materials such as Nylon or Delrin®, and the actuating wire rope *c*.

12.3.2 Stability

M-90	Unistable	KN SB

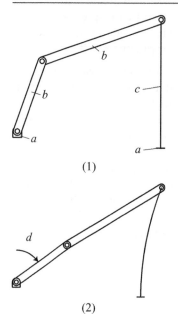

(1)

(2)

This mechanism has a cantilever beam that forces the mechanism into a single stable position when no input is applied.

(1) Rigid bodies *a* are fixed. Rigid bodies *b* are binary links. Flexible segment *c* utilizes energy transfer to hold the mechanism in this current configuration when no input is applied.

(2) Deformed (unstable) configuration of input *d*.

Bistable

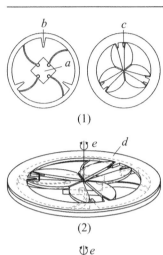

This is a multi-layer bistable mechanism. It operates from a planar configuration into a spherical configuration [59].

(1) Rigid body *a* is fixed. Rigid body segments *b* are attached to rigid segments *c*.

(2) Assembled configuration of mechanism. Rigid body *d* rotates about the *e*-axis.

(3) Deformed configuration of mechanism where rigid body *d* rotates about the *e*-axis.

(1)

(2)

(3)

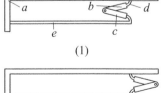

(1)

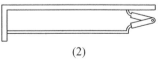

(2)

This mechanism utilizes a cantilever beam and a rigid joint to achieve bistability.

(1) Rigid body *a* is fixed. Segments *b* and *c* are rigid and segments *d* and *e* provide the flexibility.

(2) Deformed configuration and stable position.

This mechanism is a fully compliant light switch [60].

(1) Rigid body *a* is fixed. Segments *b* are living hinges and segment *c* can be modeled to produce the positions of a light switch.

(2) Deformed configuration.

(1)

(2)

This mechanism is a bistable in-plane micromechanism. A Young mechanism is defined as a mechanism with two revolute joints and two compliant segments that are part of the same link [61].

(1) Rigid segments *a* are revolute joints. The coupler segment *c* is rigid. Flexible segments *b* provide the motion and bistability.

(2) Deformed configuration and second stable position.

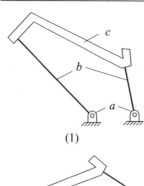

(1)

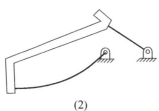

(2)

M-95   **Bistable Cylinder**

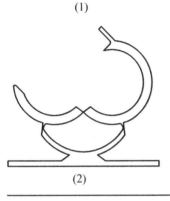

(1)

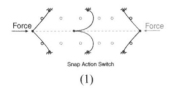

(2)

This mechanism is a clamping bistable mechanism. It opens and clamps a cylindrical object [62].

(1) Rigid body *a* is fixed. Joints *b* are living hinges.
(2) Deformed configuration and second stable position.

M-96   **Snap Action Switch**

Snap Action Switch

(1)

Microelectric switches rely on the snap action of bistable compliant mechanisms. The switch is actuated by a very small physical force. Switching happens reliably at specific and repeatable positions of the actuating point. The mechanism shown is an example of such a switch. Starting from the first stable position, the actuating force is applied at the revolute pair joining the two compliant links. The links are forced to deflect due to the physical constraints, and after a specific travel of the actuation force, the links snap to the second stable position. Similarly, an actuation force in the opposite direction snaps the links back to the first stable position.

(1) Mechanism configurations.

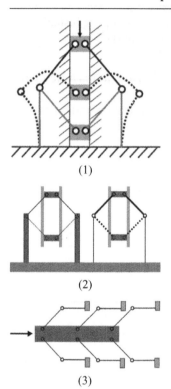

(1)

(2)

(3)

This compliant bistable mechanism is composed of a rigid slider, two flexible cantilever beams, two flexible pin-pin buckling beams mirrored at each side of the slider. One of the application areas is microswitching.

(1) Rigid slider is constrained to have a vertical motion only.
(2) The mechanism has two stable positions, i.e. when the slider is at the top and the bottom positions. It makes use of buckling of initially straight pinned-pinned beam elements and large deflection cantilever beams. Two extreme cases of the mechanism are shown: one with a rigid shoulder beam (left), the other with the rigid arm beam (right).
(3) The same topology may be used with several shoulder beams and buckling arm beams that does not require a track to follow for its shuttle (slider). Moreover, the actuation force can be increased or adjusted by adding another pair of beams.

**Partially Compliant Bistable
Six-Bar Mechanism**

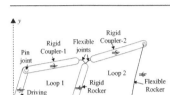

(1)

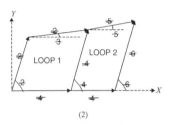

(2)

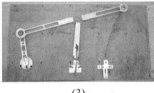

(3)

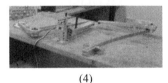

(4)

The Partially Compliant Bistable Six-Bar Mechanism consists of a rigid crank (driving link), two rigid couplers (connecting links), a rigid rocker (rocking link) and a flexible rocker (flexing link). Its pseudo-rigid-body model is equivalent to a Watt six-bar mechanism. The arrangement of the linkages makes this mechanism partly compliant. This mechanism has two stable positions where the mechanism remains motionless unless an external force or a torque is applied to displace the driving crank.

(1) Schematic diagram of the mechanism showing different linkages and the joints.
(2) Vector loop diagram of the mechanism. Here R_i, $i = 1$ through 6 denote the vectors representing the different linkages. θ denotes the angular displacement of the linkages measured counterclockwise from the horizontal.
(3) Prototype model of the mechanism showing the first stable position.
(4) Experimental setup showing the second stable position.

See also

Name	Reference Index	Categorization Index
Bistable Hinge	EM-17	FE/KN FRH/SBB
SRFBM	M-6	KM/KN TS/SBB
Bistable Planar Translator	M-7	KM/KN TS/SBB
Translational Bistable Planar	M-8	KM/KN TS/SBB
Parallel Bistable Translator	M-9	KM/KN TS/SBB
Zero-Force or Bistable Translation Mechanism	M-10	KM/KN TS/SBB
Zero Stiffness 6DoF Compliant Precision Stage	M-60	KM/KN SPP/SBB
Bistable Locking COPMM	M-62	KM/KN MM/SBB
COPMM Bistable Switch	M-63	KM/KN MM/SBB
Bistable COPMM	M-64	KM/KN MM/SBB

Multi-Stable

M-99	**Dancing Tristable**	KN SBM

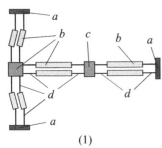

(1)

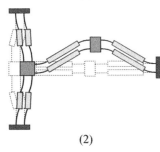

(2)

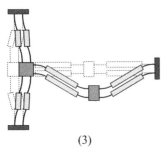

(3)

This quadri-stable mechanism, connecting a bistable mechanism with a compliant link-slider module, provides four stable equilibrium positions [63].

(1) The mechanism at its as-fabricated configuration. The motion of the end-effector c is approximately orthogonal to that of the shuttle b of the bistable part a. The rigid segments are considerably rigid compared to the flexible segments d.

(2) One of the deflected stable configurations.

(3) One of the deflected stable configurations.

M-100	**Double-Tensural Tristable Mechanism**	KN SBM

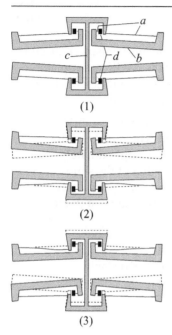

(1)

(2)

(3)

This tristable mechanism, utilizing tension flexural pivots to achieve its motion, provides three stable equilibrium positions [64].

(1) The mechanism at its as-fabricated configuration. Each of the rigid segments *b* connects two corresponding tensural pivots *a*, features *d* are fixed, and the shuttle *c* can stably stay at three distinct positions.
(2) One of the deflected stable configurations.
(3) One of the deflected stable configurations.

M-101	**Stable Core**	KN SBM

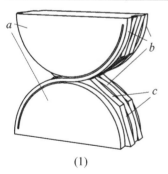

(1)

This mechanism is a compliant rolling-contact element (CORE), which connects two rigid links using flexible strips. These flexures pass between the rigid-link surfaces, and are attached to the links at the flexure ends. This element is unique such that the axis of rotation changes and is located at the contact point. This mechanism can have multiple stable positions depending on the number of stable contacts [18].

(1) Rigid bodies *a* remain in contact with each other through the flexible segments *b*. Where the axis of rotation is the contact point. Segment *c* is the stable contact.

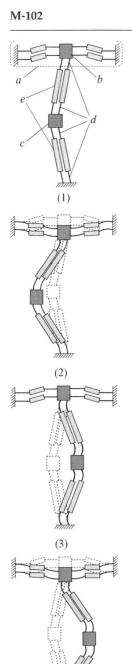

(1)

(2)

(3)

(4)

This quadri-stable mechanism, connecting a bistable mechanism with a compliant link-slider module, provides four stable equilibrium positions [65].

(1) The mechanism at its as-fabricated configuration. The motion of the end-effector c is approximately orthogonal to that of the shuttle b of the bistable part a. The rigid segments e are considerably rigid compared to the flexible segments d.
(2) One of the deflected stable configurations.
(3) One of the deflected stable configurations.
(4) One of the deflected stable configurations.

		KN
M-103	**Detent Mechanism**	**SBM**

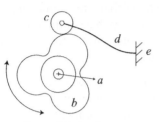

Detent Mechanism

(1)

Detent mechanism is used to hold a temporary relative position between two bodies.

(1) The body *b* rotates about a fixed axis through *a*. The roller *c* is pinned to the tip of the flexible link *d* that is fixed at *e*. The rotating body *b* has three distinct angular positions in the arrangement shown. The required torque to move from one position to another is dictated by the geometry of body *b* and the stiffness of the flexible beam. The number of holding positions is a function of the geometry of body *b*.

12.3.3 Constant Force

		KN
M-104	**Constant Force Crank Slider**	**CF**

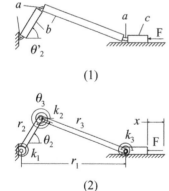

(1)

(2)

This mechanism provides a constant reaction force for a range of displacement. The mechanism configuration is a compliant slider mechanism. The mechanism achieves a constant force due to the mechanical advantage offsetting the reaction force due to deflection. Type synthesis can be performed to populate different configurations. The geometry of the rigid and compliant segments depends on the configuration [21].

(1) A permutation of a constant-force crank slider. The mechanism is composed of flexible segments *a* (small-length flexural pivots), rigid segments *b*, and a rigid slider segment *c*.
(2) Pseudo-rigid-body model.

M-105   **Electrical Connector**

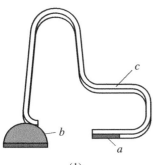

(1)

This mechanism is a constant-force electric connector (CFEC). This mechanism uses a contact cam surface and geometry to maintain a constant optimal force so fretting or adhesive wear be will less likely to occur [66].

(1) Rigid body *a* is fixed. Rigid body *b* is the cam contact surface. The flexible segment *c* is the electrical connector.

(2) Deformed configuration.

(2)

12.3.4 Force Amplification

M-106   **Pliers**

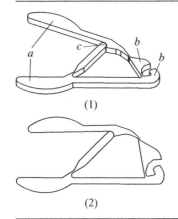

(1)

This mechanism is a fully compliant plier that, in theory, will have an infinite mechanical advantage through part of its motion [67].

(1) Rigid segments *a* are the input levers, and rigid bodies *b* are the output levers (where the force is amplified). Point *c* is a passive element.

(2) Deformed configuration.

(2)

		KN
M-107	Crimper	FA

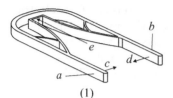

(1)

This mechanism is a fully compliant crimper. This mechanism amplifies the input force to compress an item [54].

(1) Segments *a* and *b* translate in the *c*- and *d*-direction, respectively, causing the rigid segment to deflect in the *e*-direction.

		KN
M-108	AMP Incorp. Crimpers	FA

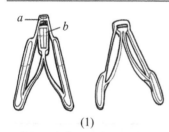

(1)

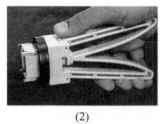

(2)

The early 1980s saw some interesting, intriguing and inspiring conceptualization and development of hand tools by AMP Incorporated, based on compliant mechanisms. Two examples shown are i) a crimping mechanism, and ii) a chip carrier extracting device.

(1) Two alternative fully compliant crimping mechanisms designed by AMP Incorporated. The mechanism is composed of the sliding segment *a*, anvil *b*, and crimped section between the slider and the anvil as the handles are squeezed.

(2) A compliant chip carrier extracting device, designed by AMP Incorporated, containing injection-molded parts. The four grasping prongs that grab the chip carrier at the corner notches and slide (extract) it out as the handles are squeezed.

KN
FA

M-109 | **Compliant MEMS Force Amplifier**

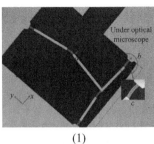

(1)

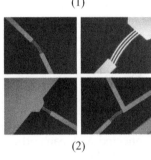

(2)

The compliant MEMS amplifier is a monolithic structure. This compliant MEMS amplifier is based on two slider-cranks attached nearly perpendicular to each other. It deflects along the x-direction by elastic beams under input force applied at point a. It provides a high output force at point b. The amplification factor, $\left(\frac{b}{a}\right)$, increases as the microcompliant micromechanism deflects along x-direction at point a.

(1) Rigid links are connected by in-plane multiple elastic links as seen in c under an optical microscope. All beams have rectangular cross-sections.
(2) Detailed views of several flexible connections of the Compliant MEMS Force Amplifier mechanism.

KN
FA

M-110 | **Compliant Crank Slider Amplifier**

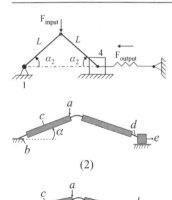

(2)

(3)

A compliant crank-slider mechanism can be used to amplify an input force.

(1) Output force and the input force ratio of the rigid slider-crank mechanism with the equal crank and the coupler lengths is: $\frac{F_{\text{output}}}{F_{\text{input}}} = \frac{1}{\tan \alpha_2}$. This ratio is called the amplification factor.
(2) The one degree of freedom compliant crank-slider mechanism shown in Figure 3 could work as force amplifier mechanism. As α gets closer to zero, $\frac{1}{\tan \alpha}$ goes to infinity. This means that a small input force applied at point a can cause a large output force at point e. Rigid body c is attached to the ground and to another rigid body by means of flexible joints, d is also an elastic joint connecting two rigid bodies.
(3) Deformed configuration of input a.

	Multistage Compliant Force Amplifier	KN
M-111	**Mechanism Design**	FA

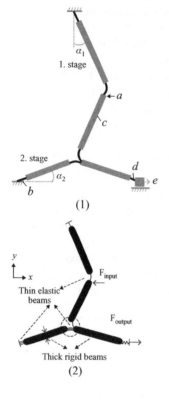

(1)

(2)

This fully compliant force-amplifier mechanism can be manufactured at the macro- and microlevels. It increases the force in each stage. The device shown has a two-stage amplification. When the force input is applied at point a; the first stage straightens and the V-shape angle α_1 gets closer to zero. As the first stage aligns with the vertical position, it pushes the lower beams down, causing α_2 to approach zero. Therefore, if the mechanism is designed such that the first stage and the second stage align with toggle positions concurrently, the output force at point e is increased double fold.

(1) The rigid bodies are fixed to the ground and connected with each other by means of elastic links.
(2) The number of stages can be increased with the addition of the same inverse V-shape topology connected to each other perpendicularly. The flexible pivots of the next stage should be made of a stack of parallel beam joints to carry the increased force transmitted to the neighboring stage.

See also

Name	Reference Index	Categorization Index
Pantograph (LEM)	M-54	KM/KN SA/FA

12.3.5 Dampening

		KN
M-112	**Dampening Ortho–Planar Spring**	DP

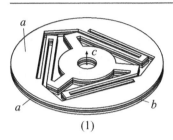

(1)

This mechanism dampens an ortho-planar spring by utilizing a viscoelastic constrained-layer for dampening. This is to reduce the free response oscillations of the spring and suppress resonance responses [68].

(1) Compliant mechanisms a are ortho-planar springs. They are separated by a viscous material, b, that allows dampening in an oscillating form in the c-direction.

12.3.6 Mode

Buckle

M-113	Partially Compliant Force-Generator Mechanism	KN MDB

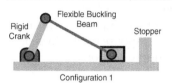

Configuration 1

Configuration 2-3

(1)

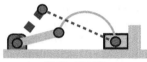

(2)

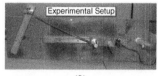

(3)

The Partially Compliant Impact and Contact Force Generator (ICFG) Mechanism is a compliant slider-crank mechanism that can be used as an impact and contact-force generator. It consists of a rigid crank, a flexible buckling beam and a stopper. The working principle of the mechanism is it behaves in two different modes: the rigid mode and the flexible mode. At the crank angle range of the mechanism, it behaves like a slider-crank. When the slider hits the stopper it generates an impact force and causes the flexible beam to buckle. The magnitude of the impact force and the contact time can be adjusted by changing the angular velocity of the actuator. The ICFG mechanism may find applications where two actions are required; a high force requirement task in a short duration (such as punching, cutting, or breaking) and a low force requirement task in longer duration (such as holding, gluing, and applying pressure) [69, 70].

(1) Top: crank-slider configuration; bottom: impact force generating configuration.
(2) Experimental setup of the crank-slider configuration.
(3) Experimental setup of the impact force generating configuration.

See also

Name	Reference Index	Categorization Index
Press	M-37	KM/KN PM/MDB

12.3.7 Others

		KN
M-114	**Force-Balance Accelerometer**	KNO

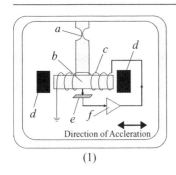

Direction of Accleration

(1)

The force-balance accelerometer measures the acceleration of a moving object and is generally used for seismic imaging, structural monitoring, and inertial navigation. The inertial force exerted on the proof mass tends to deflect the flexure hinge from its undeflected position. The deflection detector detects the motion and produces a current proportional to the acceleration, which is amplified by the signal amplifier and then fed to the coil. The interaction between the current coil and the magnet produces the counterbalancing force required to maintain the flexure hinge undeflected.

(1) The force-balance accelerometer includes a notched-type flexure hinge *a*, a proof mass *b*, a force-balance coil *c*, a magnet *d*, a deflection detector *e*, and a signal amplifier *f*.

		KN
M-115	**Piezoresistive Accelerometer**	KNO

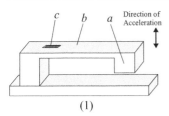

Direction of Acceleration

(1)

The piezoresistive accelerometer senses acceleration through a piezoresistive strain gauge, which proportionately responses to the deflection of the cantilever beam due to the inertial force.

(1) The piezoresistive accelerometer includes a proof mass *a*, a cantilever beam *b*, and a piezoresistive strain gauge *c*.

		KN
M-116	**Centrifugal S Clutch**	KNO

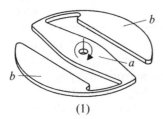

(1)

This mechanism transmits torque using centrifugal force as the actuation and control input. This mechanism can be designed to permit gradual load acceleration with a nonzero engagement speed [71, 72].

(1) Rigid segment *a* is connected to the drive shaft. Rigid segment *b* engages the device when it comes in contact.

		KN
M-117	**Plier Graspers**	KNO

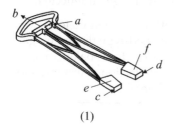

(1)

This mechanism is designed to grab items.

(1) Rigid body *a* translates in the *b*-direction, causing rigid bodies *e* and *f* to translate in the *c*- and *d*-direction, respectively.

M-118	Force-Feed Forward Mechanism for Surface Micromachined Accelerometer	KN KNO

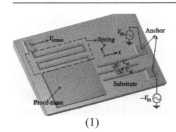

(1)

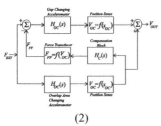

(2)

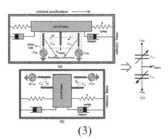

(3)

This Force-Feed Forward Mechanism increases the dynamic range of surface micromachined accelerometers. Two different types of accelerometers are used. One based on changing the gap between the comb fingers and the other changing the overlap area between the comb finger electrodes.

(1) A layout of the surface micromachined accelerometer is seen. The number of comb fingers should be made as large as possible to obtain a large sense capacitance value.

(2) Principal block diagram of Force-Feed forward method is shown. Two separate mechanical devices in this system are such that the overlap-area-changing design always forms the dynamic offset of the gap-changing design. Feedforward force from the overlap-area-changing accelerometer is applied to the proof-mass of the gap-changing design to keep it neatly catered between the fixed fingers. Because the sensitivity of the gap-changing design is much higher than the overlap-area-changing design, more precise acceleration values can be measured by the gap-changing design, while the rougher values are measured by the overlap-area-changing design.

(3) In the capacitive sense units of gap-changing accelerometer, electrostatic forces are generated on the movable parts because of the modulation voltages v_m. Because of this force, the effective spring constant of the gap changing accelerometer changes from its fixed mechanical value.

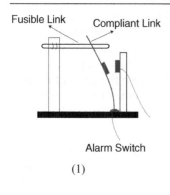

Fusible Link   Compliant Link

Alarm Switch

(1)

The alarm switch mechanism shown utilizes a compliant link to trigger the alarm.

(1) The fusible link holds the compliant link in the bent position and keeps the alarm switch open. For given specified design conditions, the strength of the fusible link deteriorates so that it cannot hold the compliant link, thus, triggering the alarm. The fusible link failure may be due to high temperature, corrosive conditions or jerky motion of the device.

References

[1] C.R. Barker, "A complete classification of 4-bar linkages," *Mechanism and Machine Theory*, vol. 20, no. 6, pp. 535–554, 1985.

[2] A. Midha, M.N. Christensen, and M.J. Erickson, "On the enumeration and synthesis of compliant mechanisms using the pseudo-rigid-body four-bar mechanism," in *Proceedings of the 5th National Applied Mechanisms and Robotics Conference*, vol. 2, 1997, pp. 93–01–93–08.

[3] N.B. Hubbard, J.W. Wittwer, J.A. Kennedy, D.L. Wilcox, and L.L. Howell, "A novel fully compliant planar linear-motion mechanism," in *ASME Design Engineering Technical Conferences, DETC04/DAC*, vol. 2 A, Sept. 2004, pp. 1–5.

[4] N.D. Masters and L.L. Howell, "A three degree-of-freedom model for self-retracting fully compliant bistable micromechanisms," *Journal of Mechanical Design, Transactions of the ASME*, vol. 127, no. 4, pp. 739–744, 2005.

[5] B.D. Jensen, M.B. Parkinson, K. Kurabayashi, L.L. Howell, and M.S. Baker, "Design optimization of a fully-compliant bistable micro-mechanism," in *ASME International Mechanical Engineering Congress and Exposition*, ser. ASME International Mechanical Engineering Congress and Exposition, vol. 2. New York, NY, United states: American Society of Mechanical Engineers, Nov. 2001, pp. 2931–2937.

[6] D.L. Wilcox and L.L. Howell, "Fully compliant tensural bistable micromechanisms (FTBM)," *Journal of Microelectromechanical Systems*, vol. 14, no. 6, pp. 1223–1235, 2005.

[7] L.L. Howell, C.M. DiBiasio, M.A. Cullinan, R. Panas, and M.L. Culpepper, "A pseudo-rigid-body model for large deflections of fixed-clamped carbon nanotubes," *Journal of Mechanisms and Robotics*, vol. 2, no. 3, 2010.

[8] N. Tolou, V.A. Henneken, and J.L. Herder, "Statically balanced compliant micro-mechanisms (Sb-MEMS): concepts and simulation," in *ASME Design Engineering Technical Conferences, DETC10/DAC*, 2010.

[9] L. Kluit, N. Tolou, and J.L. Herder, "Design of a tunable fully compliant stiffness comensation mechanism for body powered hand prosthesis," in *In proceeding of the Second International Symposium on Compliant Mechanisms CoMe2011*, 2011.

[10] J. Lassooij, N. Tolou, S. Caccavaro, G. Tortora, A. Menciassi, and J. Herder, "Laparoscopic 2DOF robotic arm with statically balanced fully compliant end effector," in *In proceeding of the Second International Symposium on Compliant Mechanisms CoMe2011*, 2011.

[11] J. Lassooij, N. Tolou, S. Caccavaro, G. Tortora, A. Menciassi, and J. Herder, "Laparoscopic 2DOF robotic arm with statically balanced fully compliant end effector," *Mechanical Sciences*, under review.

[12] N. Tolou and J.L. Herder, "Concept and modeling of a statically balanced compliant laparoscopic grasper," in *ASME Design Engineering Technical Conferences, DETC09/DAC*, 2009.

[13] J.B. Hopkins, "Design of parallel flexure systems via freedom and constraint topologies (FACT)," Master's thesis, Massachusetts Institute of Technology, Dept. of Mechanical Engineering, 2007.

[14] B.R. Cannon, T.D. Lillian, S.P. Magleby, L.L. Howell, and M.R. Linford, "A compliant end-effector for microscribing," *Precision Engineering*, vol. 29, no. 1, pp. 86–94, 2005.

[15] J.O. Jacobsen, "Fundamental components for lamina emergent mechanisms," Master's thesis, Brigham Young University, Dept. of Mechanical Engineering, 2008.

[16] B.G. Winder, S.P. Magleby, and L.L. Howell, "A study of joints suitable for lamina emergent mechanisms," in *ASME Design Engineering Technical Conferences, DETC08*, 2008.

[17] J.E. Baker, "Analysis of the bricard linkages," *Mechanism & Machine Theory*, vol. 15, no. 4, pp. 267–286, 1980.

[18] J.R. Cannon, C.P. Lusk, and L.L. Howell, "Compliant rolling-contact element mechanisms," in *ASME Design Engineering Technical Conferences, DETC05/DAC*, vol. 7 A, 2005, pp. 3–13.

[19] K. Linsenbardt, J. Mayden, J. Shaw, and J. Ward, "Design of a mechanically actuated trigger switch on a rotozip spiral saw," Missiouri University of Science and Technology, Tech. Rep., 2005.

[20] A. Tekes, U. Sonmez, and B.A. Guvenc, "Compliant folded beam suspension mechanism control for rotational dwell function generation using the state feedback linearization scheme," *Mechanism and Machine Theory*, Vol. 45, Issue 12, pp. 1924–1941, Dec 2010.

[21] L.L. Howell, *Compliant Mechanisms*. New York, NY, Wiley-Interscience, July 2001.

[22] J.M. Derderian, L.L. Howell, M.D. Murphy, S.M. Lyon, and S.D. Pack, "Compliant parallel-guiding mechanisms," in *ASME Design Engineering Technical Conferences, DETC96/DAC*, 1996.

[23] M.D. Murphy, A. Midha, and L.L. Howell, "The topological synthesis of compliant mechanisms," *Mechanism and Machine Theory*, vol. 31, no. 2, pp. 185–199, 1996.

[24] J.V. Eijk, "On the design of plate-spring mechanisms," Delft University, Netherlands, Tech. Rep., 1985.

[25] C.A. Mattson, L.L. Howell, and S.P. Magleby, "Development of commercially viable compliant mechanisms using the pseudo-rigid-body model: Case studies of parallel mechanisms," *Journal of Intelligent Material Systems and Structures*, vol. 15, no. 3, pp. 195–202, 2004.

[26] J.O. Jacobsen, B.G. Winder, L.L. Howell, and S.P. Magleby, "Lamina emergent mechanisms and their basic elements," *Journal of Mechanisms and Robotics*, vol. 2, no. 1, pp. 011 003–9, 2010.

[27] D.W. Carroll, S.P. Magleby, L.L. Howell, R.H. Todd, and C.P. Lusk, "Simplified manufacturing through a metamorhic process for compliant ortho-planar mechanisms," in *Proceedings of the ASME Design Engineering Division 2005*, vol. 118 A, 2005, pp. 389–399.

[28] R.L. Norton, *Design of Machinery*. McGraw–Hill, 2004.

[29] A.K. Saxena and N.D. Mankame, "Synthesis of path generating compliant mechanisms using intially curved frame elements," *Journal of Mechanical Design*, vol. 129, pp. 1056–1063, 2007.

[30] A.K. Rai, A.K. Saxena, and N.D. Mankame, "Unified synthesis of compact planar path-generating linkages with rigid and deformable members," *Structural and Multidisciplinary Optimization*, vol. 41, pp. 863–879, 2010.

[31] B.V. S.N. Reddy, S.V. Naik, and A. Saxena, "Systematic synthesis of large displacement contact aided monolithic compliant mechanisms," *Journal of Mechanical Design*, vol. 134, no. 1, 2012.

[32] N. Mankame and G.K. Ananthasuresh, "A novel compliant mechanism for converting reciprocating translation into enclosing curved paths," *Journal of Mechanical Design*, vol. 126, no. 4, pp. 667–672, 2004.

[33] J.O. Jacobsen, G. Chen, L.L. Howell, and S.P. Magleby, "Lamina emergent torsional (LET) joint," *Mechanism and Machine Theory*, vol. 44, no. 11, pp. 2098–2109, 2009.

[34] G. Krishnan and G.K. Ananthasuresh, "Evaluation and design of compliant displacement amplifying mechanisms for sensor applications," *Journal of Mechanical Design*, vol. 130, no. 10, pp. 102 304:1–9, 2008.

[35] H.J. Su, "Mobility analysis and synthesis of flexure mechanisms via screw algebra," in *ASME Design Engineering Technical Conferences, DETC11/DAC*, 2011.

[36] C.B. Patil, S.V. Sreenivasan, and R.G. Longoria, "Analytical and experimental characterization of parasitic motion in flexure-based selectively compliant precision mechanism," in *ASME Design Engineering Technical Conferences, DETC08/DAC*, 2008.

[37] M.L. Culpepper, G. Anderson, and P. Petri, "HexFlex: A planar mechanism for six-axis manipulation and alignment," in *Proceedings of the 17th Annual ASPE Meeting; November*, 2002.

[38] A.G. Dunning, N. Tolou, and J.L. Herder, "Design of a zero stiffness 6 DOF compliant precision stage," *under review*.

[39] G.M. Roach, S.M. Lyon, and L.L. Howell, "A compliant, over-running ratchet ad pawl clutch with centrifugal throw-out," in *ASME Design Engineering Technical Conferences, DETC98/DAC*, no. 5819, 1998.

[40] J.A. Kennedy, L.L. Howell, and W. Greenwood, "Compliant high-precision e-quintet ratcheting (CHEQR) mechanism for safety and arming devices," *Precision Engineering*, vol. 31, no. 1, pp. 13–21, 2007.

[41] J.A. Kennedy and L.L. Howell, "The rachet and pawl ring (RaPR) mechanism," in *Proceedings of the 12th IFToMM World Conference*, vol. 925, Besanco, France, June 2007.

[42] C. Gahring, "Capstone design project, mechanical engineering and applied mechanics," University of Pennsylvania, Tech. Rep., 1999.

[43] B.L. Weight, S.M. Lyon, L.L. Howell, and S.M. Wait, "Two-position micro latching mechanism requiring a single actuator," in *27th Biennial Mechanisms and Robotics Conference*, vol. 5 B, 2002, pp. 797–803.

[44] B. Hastings, T. Hendel, B. Sartin, and J. Tupper, "Design of a fluid level indicator locking mechanism," Missouri University of Science and Technology, Tech. Rep., 2008.

[45] P. Steutel, G.A. Kragten, and J.L. Herder, "Design of an underactuated finger with a monolithic structure and distributed compliance," in *ASME Design Engineering Technical Conferences, DETC10/DAC*, 2010.

[46] G. Balajia, P. Biradar, C. Saikrishna, K.V. Ramaiah, S.K. Bhaumik, A. Haruray, and G.K. Ananthasuresh, "An sma-actuated, compliant mechanism-based pipe-crawler," in *International Conference on Smart Materials, Structures, and Systems*, no. 96, July 2008.

[47] N. Mankame and G.K. Ananthasuresh, "A compliant transmission mechanism with intermittent contacts for cycle-doubling," *Journal of Mechanical Design*, vol. 129, no. 1, pp. 114–121, 2007.

[48] L. Yin, G.K. Ananthasuresh, and J. Eder, "Optimal design of a cam-flexure clamp," *Finite Elements in Analysis and Design*, vol. 40, pp. 1157–1173, 2004.

[49] T. Allred, "Compliant mechanism suspensions," Master's thesis, Brigham Young University, 2006.

[50] J.J. Parise, L.L. Howell, and S.P. Magleby, "Ortho-planar linear-motion springs," *Mechanism and Machine Theory*, vol. 36, no. 11-12, pp. 1281–1299, Nov. 2001.

[51] N.O. Rasmussen, L.L. Howell, R.H. Todd, and S.P. Magleby, "Investigation of compliant ortho-planar springs for rotational applications," in *ASME Design Engineering Technical Conferences, DETC06/DAC*, 2006.

[52] M. Dado and S. AlSadder, "The elastic spring behavior of a rhombus frame constructed from non-prismatic beams under large deflection," *International Journal of Mechanical Sciences*, vol. 48, no. 9, pp. 958–968, 2006.

[53] G.K. Ananthasuresh and L. Saggere, "A one-piece compliant stapler," University of Michigan Technical Report, Tech. Rep. UM-MEAm 95-20, Sep. 1995.

[54] G.K. Ananthasuresh and L.L. Howell, "Case studies and a note on the degrees-of-freedom in compliant mechanisms," in *ASME Design Engineering Technical Conferences, DETC96/DAC*, 1996.

[55] F.K. Byers and A. Midha, "Design of a compliant gripper mechanism," in *Proceedings of the 2nd National Applied Mechanisms and Robotics Conference*, vol. II, 1991, pp. XC.1–1 – XC.1–12.

[56] R. Adams, B. Doherty, J. Lamarr, and D. Rasch, "Force-sensing clamp," Missouri University of Science and Technology, Tech. Rep., 2010.

[57] K.V. N. Sujit, A. Saxena, A.K. Rai, and B.V. S.N. Reddy, "How to choose from a synthesized set of path-generating mechanisms," *Journal of Mechanical Design*, vol. 133, no. 9, 2011.

[58] A. Jacobi, A. Midha, et al., "A compliant gripping device," School of Mechanical Engineering, Purdue University, Tech. Rep., 1984.

[59] L.L. Howell and S.P. Magleby, "Lamina emergent mechanisms," *NSF Grant Proposal #0800606*, 2008.

[60] B.D. Jensen and L.L. Howell, "Identification of compliant pseudo-rigid-body four-link mechanism configurations resulting in bistable behavior," *Journal of Mechanical Design*, vol. 125, no. 4, pp. 701–708, Dec. 2003.

[61] B.D. Jensen, L.L. Howell, and L.G. Salmon, "Design of two-link, in-plane, bistable compliant micro-mechanisms," *Journal of Mechanical Design*, vol. 121, pp. 416–423, Sept. 1999.

[62] L.L. Howell, S.S. Rao, and A. Midha, "Reliability-based optimal design of a bistable compliant mechanism," *Journal of Mechanical Design, Transactions Of the ASME*, vol. 116, no. 4, pp. 1115–1121, 1994.

[63] G. Chen, Q.T. Aten, L.L. Howell, and B.D. Jensen, "A tristable mechanism configuration employing orthogonal compliant mechanisms," *Journal of Mechanisms and Robotics*, vol. 2, no. 014501, pp. 1–5, 2010.

[64] G. Chen, D.L. Wilcox, and L.L. Howell, "Fully compliant double tensural tristable micromechanisms (dttm)," *Journal of Micromechanics and Microengineering*, vol. 19, no. 2, 2009.

[65] G. Chen, Y. Gou, and A. Zang, "Synthesis of compliant multistable mechanisms through use of a single bistable compliant mechanism," *Journal of Mechanical Design*, vol. 133, no. 8, 2011.

[66] B.L. Weight, C.A. Mattson, S.P. Magleby, and L.L. Howell, "Configuration selection, modeling, and preliminary testing in support of constant force electrical connectors," *Journal of Electronic Packaging, Transactions of the ASME*, vol. 129, no. 3, pp. 236–246, 2007.

[67] L.L. Howell and A. Midha, "A method for the design of compliant mechanisms with small-length flexural pivots," *Journal of Mechanical Design*, vol. 116, no. 1, pp. 280–290, Mar. 1994.

[68] S. Anderson and B.D. Jensen, "Viscoelastic damping of ortho-planar springs," in *Proceedings of the ASME International Design Engineering Technical Conferences*, 2008.

[69] B. Demirel, M.T. Emirler, A. Yorukoglu, N. Koca, and U. Sonmez, "Compliant impact generator for required impact and contact force," in *ASME International Mechanical Engineering Congress and Exposition*, 2008, pp. 373–379.

[70] B. Demirel, M.T. Emirler, U. Sonmez, and A. Yorukoglu, "Semicompliant force generator mechanism design for a required impact and contact forces," *Journal of Mechanisms and Robotics*, vol. 2, no. 4, 2010.

[71] N.B. Crane, B.L. Weight, and L.L. Howell, "Investigation of compliant centrifugal clutch designs," in *ASME Design Engineering Technical Conferences, DETC01/DAC*, 2001.

[72] L.S. Suchdev and J.E. Cambell, "Self adjusting rotor for a centrifugal clutch," *United States Patent No. 4821859*, 1989.

13

Example Application

Categorized examples and descriptions of a wide range of compliant mechanisms.

13.1 Elements of Mechanisms: Flexible Elements

		FE
SM-1	**Computer Mouse**	FB

The left and right mouse buttons and the thumb button are molded plastic that deflect when pressed.

(1) The thumb button a, and the left and right mouse buttons b are made from molded plastic.

(1)

Handbook of Compliant Mechanisms, First Edition. Edited by Larry L. Howell, Spencer P. Magleby and Brian M. Olsen.
© 2013 John Wiley & Sons, Ltd. Published 2013 by John Wiley & Sons, Ltd.

		FE
SM-2	**Walker Liner Locks**	FB

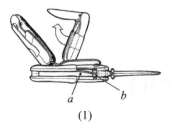

(1)

Liner locks are common in today's pocketknives. The deflection of a piece of flexible metal locks the blade of a pocketknife open. The advantage of a liner lock is that it is easy to operate with one hand.

(1) *a* is engaged so that *b* is locked open. To open the liner lock, *a* is pushed over so the blade, *b*, can close.

		FE
SM-3	**Torque Wrench**	FB

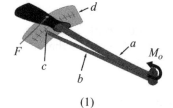

(1)

A torque wrench is used to tighten nuts to a specified torque. A long lever arm deflects due to user applied force. Another nondeflected beam points to the torque magnitude on a calibrated gauge plate.

(1) The lever arm *a* in its deflected state. The user-applied force F results in a moment M_o at the fastener. A long slender rod *b* remains undeflected because it is not in the load path. The tip *c* points to the applied torque on a calibrated plate *d*.

SM-4	Dome Tent Pole	FE FB

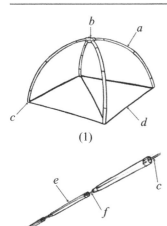

(1)

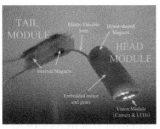

(2)

Many tents use poles to support the shape of the tent. The poles consist of a hollow cylinder with an elastic cord in the middle. The poles consist of several smaller segments. When the segments are assembled they form a compliant structure. The poles are then hooked to the tent to provide structural support for the tent walls.

(1) A schematic of the tent poles a in the assembled position. The poles are connected at the midpoint b. The bottom is held in place by inserting the pole end pin c to a grommet on the tent fabric d.

(2) Tent pole segments e are connected using a flexible cord f, and are inserted into grommets using pins c.

SM-5	Compliant Joint in a Magnetic Levitation system for an Endoscopic Camera	FE FRH

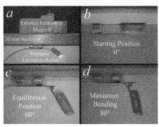

(1)

(2)

The joint in a magnetic levitation system for an endoscopic camera is designed compliant as a flexible cantilever beam. The beam deflection is due to the end-point load and moment, which are representative of the weight of the camera and the magnetic load [1, 2].

(1) Compliant joint in a magnetic levitation system for an endoscopic camera.

(2) Deformed shape for four different magnetic loads.

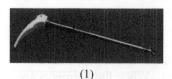

(1)

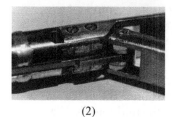

(2)

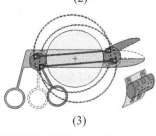

(3)

This laparoscopic instrument has high mechanical efficiency. Instead of sliding-contact pin-in-hole joints, rolling-contact surfaces were used. In order to avoid slip, flexible wrapping bands were applied. As a result, the mechanical efficiency is 96% and surgeons can perceive the pulse in an artery [3].

(1) Overview of the design.
(2) Detail of rolling-contact joint. The glossy parts are the flexible bands out of stainless steel foil.
(3) Diagram of working principle. Left: overview of two rollers on a frame; Right: detail of roller with flexible band (gray) in figure-of-eight layout.

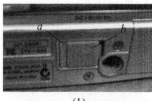

(1)

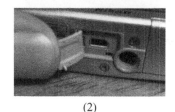

(2)

A rubber port cover protects the electronic port from dust and debris. The rubber cover can be deflected back to gain access to the electric port inside.

(1) The rubber port cover b is in its closed nondeflected state. Flexible segment a provides flexibility.
(2) The cover is deflected to gain access to the port.

		FE
SM-8	**Foldable Spoon**	FRH

(1)

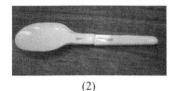

(2)

Foldable spoons use a compliant hinge to expand between compact and fully extended positions. Detents are used to keep the foldable spoon locked in position.

(1) The compliant hinge *a* allows the spoon shaft *b* to rotate and detach from the detents *c* which keep it locked in the closed position.

(2) The fully extended position of the foldable spoon.

		FE
SM-9	**Food Containers**	FRH

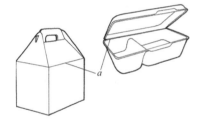

Portable food containers are used extensively to store and transport food. Flexible hinges or folds easily allow motion. Their simple design yields a low manufacturing cost and utilizes reusable or biodegradable materials.

(1) Flexible hinges or folds.

		FE
SM-10	**Battery Cover Clip**	FT

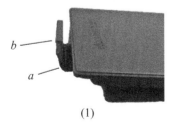

(1)

The battery cover for many commonly used devices, such as remote controls and calculators, has a flexible member and a latch that hold the cover in place. As the cover is inserted or removed, the flexible member deflects to allow the latch to move out of its locked position.

(1) The flexible member *a* has a latch *b* that allows the battery cover to lock into place.

		FE/KN
SM-11	**Slap Bracelet**	FRL/SBB

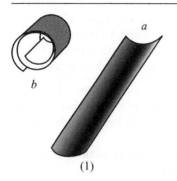

(1)

This mechanism has two stable equilibrium states. The first state occurs when it is a completely straight beam. The second state occurs when bending is applied to the straight beam and the mechanism curls up.

(1) Straight, *a*, and coiled, *b*, equilibrium states of the bracelet.

13.2 Mechanisms: Kinematic

		KM
SM-12	**Syringe Safety Cover**	TS

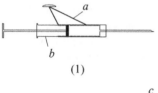

(1)

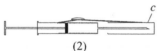

(2)

The syringe safety cover is designed to cover the sharp end of a syringe after it has been contaminated.

(1) This diagram shows the safety in a half-discharged position. The compliant segment *a* deflects to allow part *c* to slide away from the syringe body *b* covering the needle.
(2) The safety mechanism *c* in the fully discharged position and is completely covering the sharp end of the syringe.

	KM
SM-13	Wright Flier	TR

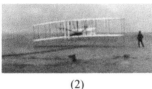

(1)

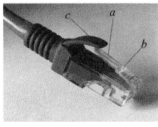

(2)

The "flying machine" developed by Orville and Wilbur Wright used wing warping for aircraft maneuver and control. The tips of the wings could be twisted using a series of cables.

(1) This drawing is from U.S. Patent No. 821,393, Flying Machine, by Orville and Wilbur Wright. The dashed lines show the deflected position of the wings.
(2) A photograph of the Wright Flier in sustained flight.

	KM
SM-14	Ethernet Cable Connector Plug	LC

(1)

(2)

The Ethernet cable connector plug is used for network cabling. The flexible tab on the plug snaps into a jack so that the plug cannot be pulled out. To get the plug out, the tab must be pressed.

(1) The flexible tab a contains a latch b that allows the plug to lock into place, and a compliant protection cap c that protects the tab from being broken.
(2) The plug is inserted and locked in a jack.

KM
LC

SM-15   **Clasp Ring**

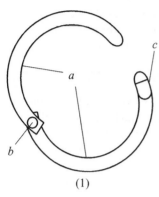

(1)

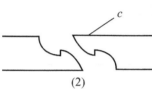

(2)

A clasp ring consists of two identical pieces of curved steel round stock.

(1) Two half circles of steel, *a*, are joined by a pin joint, *b*. In some versions of the rings the tips can flex, but usually the compliance is found along the entire member from pin to locking end.

(2) The free ends of the curves, *c*, are shaped such that they can slide together and clasp securely.

SM-16   **Blind Snap-Fit Fastener**

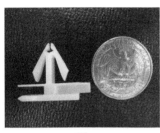

(1)

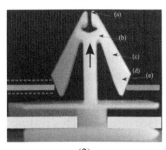

(2)

Blind snap-fit fasteners of the type shown in figure 1 are used in automotive applications for attaching components like interior trim to the sheet metal of the vehicle body. These fasteners allow trim pieces to be attached quickly and with low effort. Unlike screws and bolts, these are nonreversible, i.e. the attachment cannot be undone without damaging the trim and/or the sheet metal. They are completely hidden between the trim and the sheet metal (hence, blind) and they take up less space than conventional fasteners. The fastener is inserted into a housing in the trim piece (shown in figure 2). The nose a of the fastener is lined up with a hole in the sheet metal and the assembly comprising the trim and the fastener is pressed into the hole in the direction shown by the arrow. The tapered nose a provides a self-centering action. Two arms are cantilevered off the central post by flexures b. During the assembly process, the arms deflect towards the central post as the fastener moves into the sheet metal. The maximum deflection occurs at the edge d. When the sheet metal passes the edge d, an audible click registers the completion of the assembly process. The profile of the arms changes to e when the sheet metal moves beyond d. This causes the arms to spring back and re-engage the sheet metal. The profile of the region e allows the fastener to self-center in the hole in the sheet metal while accommodating variations in the hole diameter. The dashed lines indicate the limiting positions of the sheet metal for a range of admissible hole sizes.

(1) A nonstructural, blind, snap-fit fastener used to attach trim pieces to the sheet metal body of an automobile
(2) Schematic showing the fastener in its assembled condition.

(1)

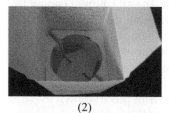

(2)

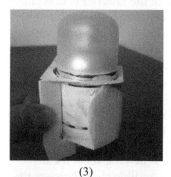

(3)

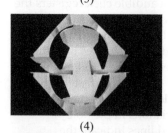

(4)

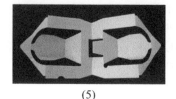

(5)

A packaging insert made of corrugated card paper is used to protect the fragile contents of a perfume container (figure 1). The insert is folded to create a three dimensional nest for a perfume bottle (figures 2 and 3). The container provides form closure to constrain and support the folded configuration of the insert. When the constraints are removed, the elastic energy stored in the mechanism causes it to unfold partially (figure 4). The intricate geometry and the various interlocking segments of the insert are visible in this figure. The insert exemplifies an ortho-planar compliant mechanism which can be rearranged so that all of its links lie in a single plane (see figure 5). This feature allows ortho-planar compliant mechanisms to be fabricated in a single operation in their planar condition. They are subsequently folded into three-dimensional configurations to perform the desired tasks as in the case of the packaging insert described here.

(1) Top view of the opened package shows the bottle nestled in the packaging insert.
(2) A top view without the bottle shows the folded configuration of the insert.
(3) The insert and the bottle outside the carton.
(4) The insert in a partially unfolded configuration.
(5) The insert in a nearly completely flattened configuration.

SM-18   **Bourdon Tube Pressure Gauge**

(1)

(2)

A Bourdon tube is an instrument used to measure the pressure of a liquid or gas. It has a flexible curved tube that can be filled with fluid. As the pressure inside the device increases, the tube straightens. The deflection of the tube is used to measure the pressure of the fluid inside.

(1) Pressure inlet *a* allows fluid to enter curved tube *b* with cross section *c*. As the pressure increases, tube *b* deflects outward. This deflection is transferred by link *d* to the measurement readout system. Gear sector *e* rotates around fixed pivot *f*. The teeth on gear sector *e* cause rotation of pinion *g* and pointer *h* where the current fluid pressure can be read.

(2) A photograph of a Bourdon tube pressure gauge.

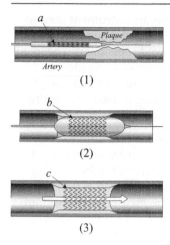

(1)

(2)

(3)

A coronary stent is a wire tubular structure that is deployed to unclog a heart artery. It remains in the artery after an angioplasty, or percutaneous coronary intervention (PCI) to help it remain open. It is deployed with the help of a balloon catheter that inflates and expands the wire tube to a desired diameter that will increase blood flow to the heart. (Aside: the balloon catheter itself is also a compliant mechanism and can be used by itself to unclog an artery.) Each wire is a compliant mechanism and can be modeled as a pinned-pinned segment.

(1) The coronary stent, *a*, in its undeflected position. The stent is positioned over the balloon catheter, which is also in its undeflected position.
(2) The inflated balloon catheter helps remove the blockage, usually through a series of inflations and deflations in a deflected position. The expansion of the stent and balloon catheter is marked *b*.
(3) The stent at the end of deployment providing structural support to the artery to improve blood flow.

SM-20   **Drywall Mount**

(1)

(2)

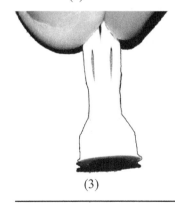

(3)

The drywall mount is designed to distribute the force of a hanging object over a large area of the drywall. By dispersing weight over a larger area, drywall can support more weight. The mount is closed and inserted into a predrilled hole. The mount then opens to disperse the weight and allow a hook to be secured into it.

(1) The drywall mount in the open position. This allows the force of the hanging object to be dispersed over a larger area. The compliant segments a, b, and c bend to allow the mount to enter the closed position.
(2) The drywall mount in the partially closed position.
(3) The drywall mount in the completely closed position. This position requires the user to input force.

KM

SM-21   **Zipper**   KMO

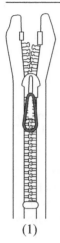

(1)

The zipper consists of a series of teeth on two sides, driven by a slide in the middle. The teeth are slightly compliant, and the mounting fabric behind the teeth is very compliant. The tip of each tooth is wider and rounded. This is called the hook. Behind the hook on each tooth is a hollow. The slide brings the hooks past each other at an angle to position them in the hollow of the tooth across from them. In this way the teeth are firmly held together.

(1) The slide is movable by the pull to open and close the teeth.

KM

SM-22   **Extendable Handle**   KMO

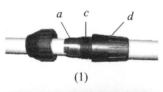

(1)

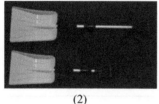

(2)

The extendable handle shaft allows a handle to adjust its length. The two parts of the handle shaft have different diameters and the smaller part can slide into the larger part. A compliant collet is wed to clamp the two parts in position.

(1) As the threaded halves *a* and *b* are twisted together the top half, *a* applies a force normal to the compliant collet, *c*. The compliant members are pressed inward until forced against the inner shaft, holding the outer shaft and inner shaft in the desired position.
(2) The handle extension can be used in multiple applications, including this shovel.

13.3 Mechanisms: Kinetic

		KN
SM-23	Chair	ES

a

b

(1)

This traditional-looking chair is made of a cushioned seat suspended on a U-shaped wooden frame. The compliant wooden frame bends to provide a slight deflection for comfort. The flexible frame allows a person to rock or bounce.

(1) The cushioned seat *a* is supported by the bent beechwood layer-glued frame *b* that provides deflection downward as indicated by the arrow.

Reproduced by permission of IKEA

		KN
SM-24	Atomic Force Microscope	ES

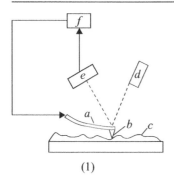

(1)

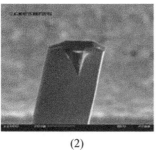

(2)

An atomic force microscope uses a sharp tip connected to a flexible cantilever to scan the surface of a specimen. The cantilever deflects as it goes over the surface, and its motion is detected and the resulting force can be calculated using Hooke's law.

(1) The cantilever *a* has a tip *b* that goes across the surface *c*. The system also includes a laser *d*, photodiode *e*, and detector and feedback electronics *f*.
(2) A scanning electron micrograph of an AFM cantilever tip.

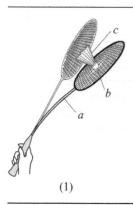

(1)

Upon impact with a shuttlecock, the flexible shaft and the tight string of a badminton racket store energy as they deflect and then deliver the energy to the shuttlecock.

(1) A badminton racket both in its deflected and undeflected states. The flexible shaft *a*, the string *b*, and the shuttlecock *c*.

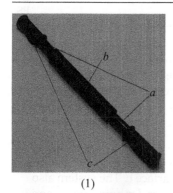

(1)

(2)

The power twister bar is a home fitness tool for building upper-body strength. The power twister is a bar with a spring coil in the middle and rubber handles on the ends.

(1) A power twister bar includes two rigid segments *a*, a spring coil *b*, and two rubber handles *c*.

(2) A power twister bar in a deflected position.

SM-27   **Statically Balanced Gripper**

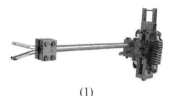

(1)

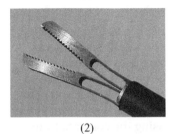

(2)

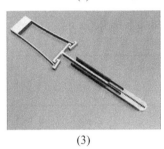

(3)

This design concerns a surgical gripper with a fully compliant gripper. Despite the advantages due to the compliant mechanism, the stiffness due to deformation distorts the force feedback for the surgeon. To eliminate this problem, a negative stiffness mechanism (balancer) was added to cancel out the stiffness. As a result, the mechanical efficiency is improved and force feedback is restored [4].

(1) Overview of the instrument with partly compliant balancer.
(2) Close-up of the fully compliant gripper.
(3) Fully compliant design.

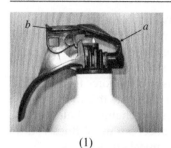

(1)

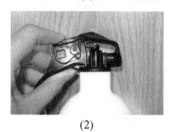

(2)

A fire extinguisher uses compressed dry chemical to suppress a growing fire. Inside the main chamber is a cartridge filled with CO_2. Depression of the handle causes a pin to puncher the cartridge, increasing the pressure within the canister and forcing the contents out of the nozzle.

(1) Rather than using a spring positioned between two handles, this particular fire extinguisher uses the flexibility of the material at point a to allow the handle, b, to depress the pin and open the valve that increases pressure in the canister, releasing the contents from the nozzle.

(2) The extinguisher handle in its deflected potion.

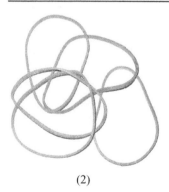

(2)

The rubber band, a common household item, is a simple compliant object that utilizes the elasticity of rubber. It can be used as a compliant structure to hold objects together. It can also be used as a compliant mechanism to return energy to a system when initially stretched.

(1) Rubber band.

	KN
SM-30	Nail Clippers	**ES**

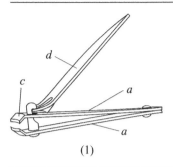

(1)

Traditional nail clippers use compliant members to allow motion and provide spring back.

(1) Nail clippers use two compliant members, *a*, connected to the jaws, *b*. The jaws are at the end of the cantilevered beams. The compliant beams deflect when the lever, *c*, is activated.

	KN
SM-31	Moose Bottle Nozzle	**ES**

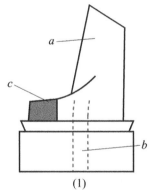

(1)

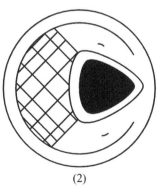

(2)

The nozzle cover provides the user with a comfortable way to dispense hair product from the pressurized mousse container while causing the product to foam when it exits the nozzle. The cover deflects under force from the user, releasing the pressurized product. As the product exits, small flexible plastic veins cut the stream, promoting a foaming effect. When the user lets up on the cover, product ceases to dispense.

(1) A side view of the undeflected nozzle cover *a*. The shaded surface *c* is where the force is applied. Segment *b* is the nozzle.

(2) A top view of the nozzle cover showing the surface, *c*, where force is applied, and compliant segment, *b*, which undergoes bending when a force is applied.

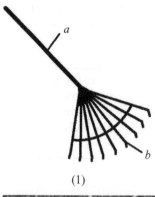

(1)

(2)

The rake is a common agricultural tool used for yard work. As the rake is pulled across the ground, its flexible teeth deflect up and down, keeping the rake in full contact with the raking surface.

(1) The rake consists of a rigid pole *a* that is attached to a series of compliant teeth *b*, typically constructed of plastic or metal.
(2) Rakes come in various shapes and sizes, some with potential to deflect more than others.

(1)

(2)

The Slinky® is made from a single flexible metal (or plastic) beam formed into a cylindrical shape. It stores energy as it is extended from its equilibrium state.

(1) Undeflected position.
(2) The Slinky® in a 'stepping' form, wanting to return to the undeflected state.

SM-34   Wire Gate Carabiner

(1)

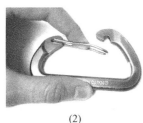

(2)

This "wire gate" carabiner is used in combination with a "straight gate" carabiner in sport climbing. The straight gate design consists of a pin joint, a latch, a gate and a spring. The wire gate design combines all the straight gate components into one compliant member. This allows for a lighter, safer and more compact design while climbing.

(1) The wire gate *a* compared to the standard straight gate *b*. Both are in the closed position.
(2) A force is applied on the wire gate. The gate is displaced and the carabiner is open. Upon releasing, the gate will spring back into the closed position.

SM-35   Negative Pressure Pump

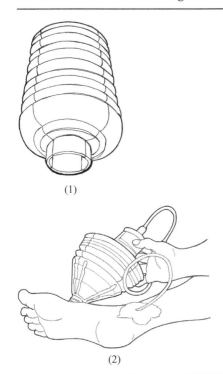

(1)

(2)

Applying negative pressure to a wound can reduce healing time, especially when more advanced treatments, such as stitches, are not available. Pumps to apply suction to treated wounds are expensive and require batteries to operate, but MIT developed a compliant hand pump to assist third-world countries in times of natural disaster. The device creates suction as a user presses the plastic accordion folds together. It is made from a tube and an injection-molded polypropylene casing, making it a fraction of the cost of traditional pumps.

(1) Uncompressed pump.
(2) Compressed pump.

SM-36   **Keyboard Key**

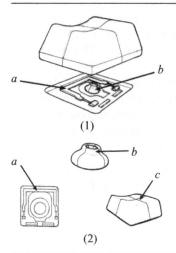

(1)

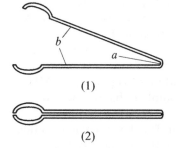

(2)

Underneath each keyboard key is a silicone or rubber nipple that provides the spring needed to return the key after being pressed. The plastic piece keeps the key level during its motion.

(1) A picture of a keyboard with one key top piece missing.
(2) Components of a keyboard key assembly: the support piece *a* keeps the key level, the rubber nipple *b* used to provide a reaction force, and the visible key *c*.

SM-37   **One-Piece Tongs**

(1)

(2)

One-piece tongs use a single plastic or metal flexible segment to allow clamping motion and provide return force.

(1) The tongs in open position. Flexible segment *a* provides spring back energy storage and allows the clamping motion. The rigid segments *b* do not deflect when grabbing objects.
(2) The tongs in closed position.

SM-38   **Key Ring**

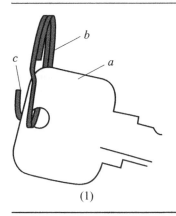

(1)

A common key ring is designed with a metal wire.

(1) To put a key *a* on the ring *b*, one portion of the metal is deflected outward *c*.

SM-39   **Pin Clutch**

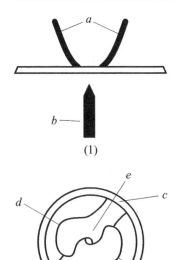

(1)

(2)

The pin clutch is used to grasp the back of a pin and hold it in place. The metal fins are pushed aside by the inserted pin and then clasp onto it. When the fins are pressed together, it releases the clutch. These are used with tie pins, label pins, and decorative pins.

(1) Side view of clutch and pin. Metal fins, *a*, that are pressed to release the clutch pin, *b*. The metal arch, not shown, does not affect the compliant mechanism.

(2) Top view of pin clutch without metal arch over the middle. The points of deflection of metal fins, *c*, and flat sections of metal pins, *d*, which come up at an angle. The pin needle is inserted through hole, *e*. The hole diameter increases when the metal fins are pushed together and decreases when released, grasping the pin.

KN
SM-40   **Sling Shot**   ES

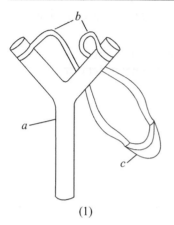

(1)

Slingshots use flexible elastic bands that store and release energy to launch projectiles.

(1) A rigid segment *a* provides support and a handle. The flexible segment *b* provides energy storage when actuated. This flexible segment *c* interacts with *b* and holds, and releases the projectile.

KN
SM-41   **Bow**   ESC

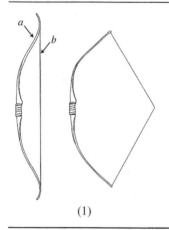

(1)

A bow is made of flexible material that allows the limbs to be deflected when the string is pulled. When the string is released the energy stored in the deflected members is transformed into the kinetic energy of the arrow.

(1) Left: Undeflected position of the bow. Right: Deflected position of the bow. The limbs *a* deflect when the string *b* is drawn.

SM-42   **Swimming Goggles**

(1)

The swimming goggles are used to keep water from getting into swimmers' eyes. They have an adjustable elastic head-strap and a flexible nosebridge so that they can fit different faces.

(1) The main components of swimming goggles include: a nosebridge *a*, a head-strap *b*, and glass cups *c*.

SM-43   **Sack Clip**

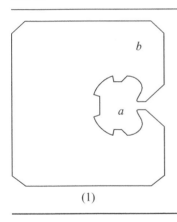

(1)

A plastic sack can be held closed with a sack clip.

(1) A plastic sack is pushed into the sack slot *a* and held into place by the hooks *b*. The clip can be deformed out of plain to release the sack. The compliant clip can be reused.

SM-44   **Clip**   KN
                                                                                             ESC

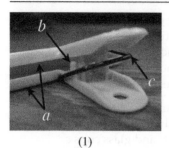

(1)

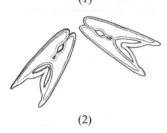

(2)

The clip uses a bent metal wire as a spring to hold objects between plastic clamps. The clamping action holds opened food bags shut and food fresh. Other clips are monolithic, where the compliant member is integrated into the clip design.

(1) Two plastic pieces, *a*, are connected by a rocker joint *b* and are held together by the force exerted by the metal wire *c*. This keeps the parts aligned when the clip is opened, causing the metal wire to deflect and applying a force at the end when the pieces have an object placed between them.

(2) Monolithic clips.

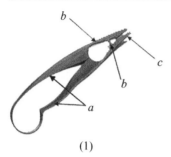

(1)

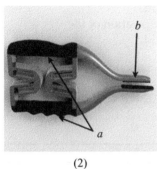

(2)

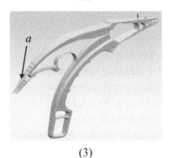

(3)

Compliers®, a fish hook remover, is a one-piece compliant tool composed of flexible and rigid segments. When the handles are squeezed, the flexible segments deform giving the mechanism its motion, while causing the jaws to exert a grasping force [5, 6].

(1) Compliers®, showing handles a, flexural pivots b, and jaws c. Hook removal uses grasping and disgorging actions.

(2) This design allows parallel and nonparallel articulation of the jaws b, depending on where the input forces are applied on the handles a.

(3) This design features a pair of hook removers in one, the smaller one a at rear is for a fly fishing application. The jaws are specialized to grasp the hook in various modes.

KN
SM-46   **Bistable Compliant Sippy-Cup Lid**   SBB

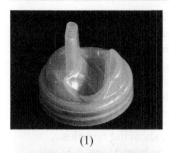

(1)

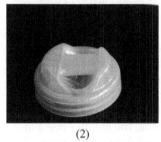

(2)

The threaded lid of the sippy-cup is a single-piece structure with a flexible shell of suitable shape that provides bistable behavior to the protruding part in the first figure. Liquid inside the cup can be sipped in this configuration. Leaking is prevented when folded as shown in the second figure.

(1) Open stable position of the sippy-cup's lid.
(2) Closed stable position of the sippy-cup's lid.

KN
SM-47   **Compliant Off-Shore Leg Platform**   SBB

(1)

The figure shows how an off-shore platform is held by flexible wires kept in tension by a pontoon supporting the columns of the platform. The tension in the wires is adjusted to stabilize the platform. This design is similar to marine plants whose roots are on the sea-bed while they stay afloat.

(1) Schematic of the compliant leg platform.

		KN
SM-48	**Foldable Bucket**	SBM

(1)

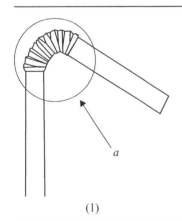

(2)

This foldable bucket consists of a number of bistable structures connected serially, thus exhibiting a multistable behavior.

(1) The foldable bucket is fully extended.
(2) The foldable bucket in its collapsed state.

		KN
SM-49	**Flexible Straw**	SBM

The flexible straw is made from one piece of plastic. A corrugated section allows the end of the straw to be positioned to the desired angle.

(1) The corrugated section *a* allows the end of the straw to move, but also holds it in place.

a

(1)

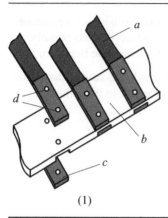

(1)

Potato and sugar beet harvesters, and many other types of equipment, use belted chains as a means of conveying of produce while allowing dirt, water or other particles to drop through. Reliability and longevity of the belted chain are superior to traditional approaches, such as hook link chains.

(1) Belted Chain construction, using rigid steel link *a* connected to flexible rubber belting *b* using steel backing plates *c* and steel rivets *d*.

(1)

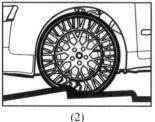

(2)

This airless wheel has a shock-absorbing rubber tread band that distributes pressure to flexible polyurethane spokes supported by an aluminum wheel. It has advantages of being maintenance-free, puncture-proof, easy to mount and dismount, easy to retread, and provides longer service than radial tires.

(1) The tire tread *a* is connected to the deformable wheel *b* and flexible spokes *c*.
(2) A deformed position of the tire.

SM-52   **Suction Cup**

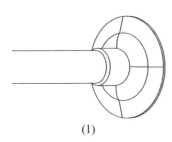

(1)

A suction cup uses negative fluid pressure of air to adhere to smooth and nonporous surfaces. It is often used to affix light objects to nonporous vertical surfaces such as windows, refrigerator doors and tiled walls.

(1) Suction cup.

SM-53   **Blood Pressure Cup**

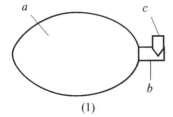

(1)

A blood pressure pump is a hand-operated device that creates pressure by squeezing the pump. The device is connected by a hose to a cuff that wraps around your arm, to measure blood pressure.

(1) The flexible surface *a* compresses and forces air through the tube *b*. The relief valve *c* allows the air to pass one way; it twists to allow the air to escape.

SM-54   **Automotive Seal**

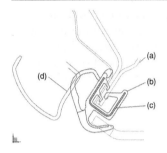

Seals located between the doors and the body of an automobile play an important role in the performance of the vehicle as well as its perceived quality. They keep undesirable elements like water, snow, dust and external noise out of the passenger compartment. They also serve to isolate the climate-controlled passenger compartment from the exterior environment. A body-mounted secondary door seal comprises an elastomeric body b mounted on a metal stiffener c. It sits between the automobile body a and the door d. The seal gets compressed when the door is closed. Good sealing function can be achieved by a seal that is flexible enough to conform to the door surface and stiff enough to maintain a sufficient contact or sealing pressure. However, a stiff seal requires users to exert a lot of effort to close the door and hence is not desirable from an ergonomic viewpoint. Perceptual quality studies have shown that stiff seals also lead to poor sound quality during the door-closing process. Hence, seal designers seek to balance these conflicting functional requirements by a proper choice of geometry and material. Modern vehicles, typically, have a sealing system that comprises multiple seals (e.g. primary, secondary, auxiliary) to achieve an optimum balance between the sealing and other functional requirements.

(1) A body-mounted secondary door seal in its as-manufactured configuration shown in its proper location on the vehicle.

SM-55   **Blood Vessel**

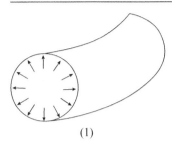

(1)

Blood vessels expand and contract circumferentially as cylindrically elastic tubes. The contraction and expansion of a blood vessel are called vasoconstriction and vasodilation, respectively. After contraction of the left ventricle, the arteries expand to reduce the change in pressure (explaining why diastolic pressure is lower than systolic, but not zero).

(1) The smooth muscle along the walls of the blood vessel enable vasoconstriction and vasodilation.

SM-56   **Muscle Fiber**

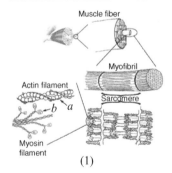

(1)

Muscle fibers provide a longitudinal contraction by the movement of the myosin heads b along the actin filaments a. When the muscle is relaxed, the myosin heads detach from the actin filament and the contraction is released.

(1) The figure shows the relationship of the actin a and myosin b filaments to the muscle fiber. The actin and myosin filaments provide the contractile motion that is characteristic of all types of muscle (skeletal, cardiac, and smooth).

SM-57   **Proteins**

Proteins self-assemble into their folded structure, which seems to be dependent only on its sequence to define the native fold (alpha helix or beta sheet). The folded structure represents the lowest-energy state for that protein (with a few exceptions).

(1) The figure shows an illustration of a murine anticholera antibody with bound carbohydrate antigen. The two protein chains are colored blue and orange.

(1)

SM-58   **Heart Valves**

The four heart valves are composed of two or three collagen membrane leaflets that allow unidirectional flow. The leaflets undergo bending loads in the direction of flow. Deceleration of flow causes a positive pressure gradient that closes the valve.

(1) The pulmonary, aortic, bicuspid, and tricuspid heart valves are labeled in the figure. The flexibility of the collagen membrane enables the leaflets to open and close.

(1)

KN
SM-59   **Compliant Heart-Valves Insert**   KNO

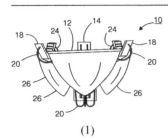

(1)

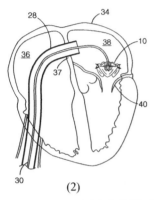

(2)

The percutaneous heart valve shown in the figures consists of a bistable compliant mechanism [8]. It has a flexible ring on which three valve-leaves are attached with a central compliant segment, which can be pulled to deploy or fold the valve reversibly [7].

(1) Heart valve.
(2) Heart valve being inserted into the heart.

KN
SM-60   **Esophagus**   KNO

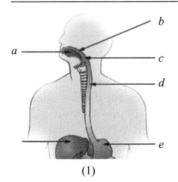

(1)

The esophagus is a cylindrically elastic tube that uses serial contraction (peristalsis) to convey food from the mouth to the stomach.

(1) The esophagus forms part of the digestive tract, connecting the mouth to the stomach. The parts of the digestive tract pictured are: *a* tongue, *b* mouth, *c* pharynx, *d* esophagus, and *e* stomach.

		KN
SM-61	Iris	KNO

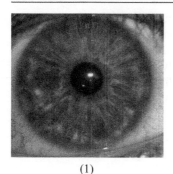

(1)

The iris is a thin, contractile membrane that controls the amount of light that enters the eye. The iris is also what determines eye color.

(1) An example of an iris is shown in the figure.

		KN
SM-62	Lens with Ciliary Muscles	KNO

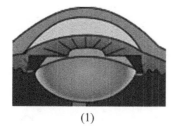

(1)

The lens is primarily composed of fibers within an elastic collagen capsule. The lens focuses light on the retina by changing shape (rounder or flatter), which movement is controlled by the ciliary muscles on the periphery.

(1) The biconvex lens is shown.

		KN
SM-63	Erythrocytes (Red Blood Cells)	KNO

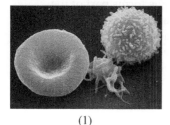

(1)

Erythrocytes are biconcave disks that can nearly double their volume without membrane distension. When passing through narrow blood vessels, the erythrocytes release ATP to relax the vessel walls.

(1) A scanning electron micrograph of an erythrocyte.

SM-64	Young Plant Stems	KN KNO

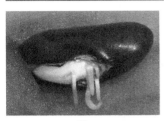

(1)

(2)

Young plants rely on the flexibility and compliance of their stems to break through the ground.

(1) The stems grow out of the seed with a bent shape to make their way through the ground.
(2) While they grow, the stems form loops under the ground storing enough strain energy at certain conditions to overcome the soil resistance. They pop out at the right moment and push the top soil layer out. Usually, this happens when the ground is wet.

SM-65	Corn-Stack Compliant and Strong Design	KN KNO

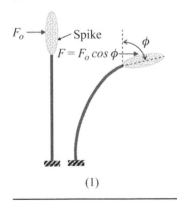

F_o Spike

$F = F_o \cos \phi$ ϕ

(1)

The stems of cereal crops possess a remarkable compliant design to prevent their uprooting under strong wind-loads. By flexing their stems, they not only reduce the drag loads by aligning the grain-bearing spike but also reduce the uprooting reaction moment at the ground level. Leaves, twigs, and branches of plants and trees effectively use their compliance in this manner [8].

(1) Corn-stack's compliant design.

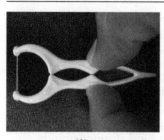

(1)

(2)

A disposable dental floss pick exhibits the key functional elements of contact-aided compliant mechanisms. The two symmetric halves of the floss pick are connected at three locations: via an inextensible floss segment at the left end of a via b compliant segment that ends in a toothpick and via c a flexure joint in between these two ends. The user grasps the floss pick and presses its sides as shown in figure 2. This causes the two halves of the floss pick to pivot about the flexure joints thereby tensioning the floss. Tensioning the floss allows it to slide easily between teeth. If the user presses too hard, the two halves of the floss pick come into contact as seen in figure (2). This contact interaction prevents any further tensioning of the floss. After the floss is slid between two teeth, the user reduces the grip pressure on the sides of the floss pick to relieve the tension in the floss. This allows the floss to conform to the tooth surface and thus cover more surface per flossing stroke.

(1) A disposable floss pick uses compliance and unilateral constraints to achieve greater functionality than conventional floss picks.
(2) Contact and tension-only (string) constraints play a key role in the function of the floss pick.

SM-67   **Centrifugal Clutch**

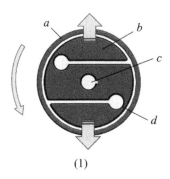

(1)

A compliant centrifugal clutch utilizes centrifugal force to transfer motion from an input rotational shaft to an outer drum, which is connected to an output shaft. Centrifugal clutches allow the input shaft to rotate without rotation of the output shaft (such as an idling engine), but engage the output shaft at a given speed.

(1) The outer drum *a* makes contact with arm *b* once the rotation from shaft *c* becomes high enough to deflect the flexible segment *d*. The mechanism has flexible arms on both sides.

SM-68   **Piston Cup Seal**

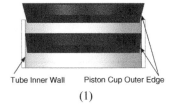

Tube Inner Wall   Piston Cup Outer Edge

(1)

A piston cup uses a flexible edge for sealing in hydraulic applications. For example, when used in a water pump, the piston cup allows for the pressure needed for the pump to perform its function.

(1) The piston cup has a larger initial diameter than that of the interfacing tube. When placed into the tube, the outer edge flexes against the tube's inner wall, providing the seal.

SM-69   **Bulb Syringe**

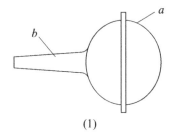

(1)

A bulb syringe is a hand-held device that is used to extract mucus from an infant's mouth and lungs. The bulb is squeezed and then released to form a vacuum.

(1) The flexible surface *a* is compressed and then released to pull fluids through the tube *b*.

SM-70   **Retaining Ring**   KN
KNO

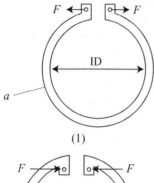

(1)

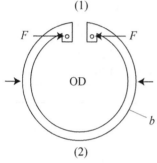

(2)

A retaining ring is used to prevent shafts from sliding along their rotational axis without constraining their rotation. While they do not deflect during normal use, they must be deflected during assembly.

(1) The external retaining ring a is loaded with forces F on the end which expands the inner diameter (ID) allowing it to slide over the outside of the shaft to which it is installed.

(2) The internal retaining ring b is compressed so that its outer diameter (OD) fits inside the hollow shaft to which it is installed.

SM-71   **Ice Cube Tray**   KN
KNO

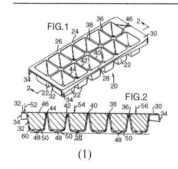

(1)

To remove ice from the plastic enclosure, a user twists the two ends in opposite directions. The resulting deflection releases the ice from multiple tapered rectangles. Previously, ice cube trays were made from aluminum and were not compliant

(1) Figure of flexible ice tray.

		KN
SM-72	**Rubber Track**	KNO

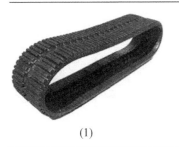

Some off-road equipment uses rubber tracks that provide good traction and are easy to install. Rubber tracks are usually used on small equipment.

(1) Rubber track.

(1)

Reproduced by permission of McLaren Industries, Inc.

References

[1] M. Simi, N. Tolou, P. Valdastri, and J. L. Herder, "Modeling of a compliant joint in a magnetic levitation system for an endoscopic camera," in *In Proceedings of the Second International Symposium on Compliant Mechanisms CoMe2011*, 2011.

[2] M. Simi, N. Tolou, P. Valdastri, J. L. Herder, A. Menciassi, and P. Dario, "Modeling of a compliant joint in a magnetic levitation system for an endoscopic camera," *Mechanical Sciences*, under review.

[3] J. L. Herder, M. J. Horward, and W. Sjoerdsma, "A laparoscopic grasper with force perception," *Minimally Invasive Therapy and Allied Technologies*, vol. 6, no. 4, pp. 279–286, 1997.

[4] J. L. Herder and F. van den Berg, "Statically balanced compliant mechanisms (SBCM's), an example and prospects," in *ASME Design Engineering Technical Conferences, DETC00/DAC*, 2000.

[5] (2001). [Online]. Available: http://compliersinc.com/

[6] S. Oswald, T. Hilse, P. Stieff, R. Trulli, D. Ems, O. Johns, K. Giovannini, T. Tran, J. Ming, and J. Michael, "Design of a fish hook remover," Missouri University of Science and Technology, Tech. Rep., 2011.

[7] H. C. Hermann, N. Mankame, and G. K. Ananthasuresh, "Percutaneous heart valve," *United States patent US 7,621,948 B2*, Nov. 2009.

[8] P. Sivanagendra and G. Ananthasuresh, "Size-optimization of a cantilever beamunder the deformation dependent load with application to wheat plants," *Structural and Multidisciplinary Optimization*, vol. 39, no. DOI 10.1007/s00158-008-0342-4, pp. 327–336, 2009.

Index

Note: Bold page references relate to images in Part IV.

Handbook of Compliant Mechanisms, First Edition. Edited by Larry L. Howell, Spencer P. Magleby and Brian M. Olsen.
© 2013 John Wiley & Sons, Ltd. Published 2013 by John Wiley & Sons, Ltd.

Printed and bound by CPI Group (UK) Ltd, Croydon, CR0 4YY

16/04/2025

14658554-0005